James Brodie Gresswell, Albert Gresswell, George Gresswell

Diseases and Disorders of the Horse

A Treatise on Equine Medicine and Surgery

James Brodie Gresswell, Albert Gresswell, George Gresswell

Diseases and Disorders of the Horse
A Treatise on Equine Medicine and Surgery

ISBN/EAN: 9783337778392

Printed in Europe, USA, Canada, Australia, Japan

Cover: Foto ©berggeist007 / pixelio.de

More available books at **www.hansebooks.com**

DISEASES & DISORDERS

OF THE

H O R S E :

A Treatise on Equine Medicine and Surgery,

BEING A CONTRIBUTION TO THE SCIENCE OF COMPARATIVE PATHOLOGY,

BY

ALBERT GRESSWELL,

Graduate in High Honours, and late Junior Student of Christ Church, Oxford ; Member of the
Royal College of Surgeons of England ; Author, in conjunction with Mr. J. B. GRESSWELL,
of the " Manual of the Theory and Practice of Equine Medicine ; " and of
" The Equine Hospital Prescriber," &c.;

AND

JAMES BRODIE GRESSWELL, M.R.C.V.S.,

Author of the "Veterinary Pharmacology and Therapeutics," and other Works and Papers on
Veterinary Science ; Veterinary Inspector for the Lindsey Division, and for the Borough
of Louth, Lincolnshire ; Provincial Veterinary Surgeon to the Royal
Agricultural Society ;

REVISED, WITH AN INTRODUCTION BY

GEORGE GRESSWELL,

Graduate in Honours, and late Open College Exhibitioner of Christ Church, Oxford ; Graduate
of the University of the Cape of Good Hope ; Author of the " Evolution Hypothesis,"
" The Wonderland of Evolution," " The Place of Physical Science in
Education," &c.; recently Lecturer in Physical Science under the
Government of the Cape of Good Hope.

LEEDS:

PUBLISHED BY THE YORKSHIRE CONSERVATIVE NEWSPAPER CO., LIMITED,

1886.

Sed quum tota philosophia, mi Cicero, frugifera et fructuosa, nec ulla pars ejus inculta ac deserta sit, tum nullus feracior in ea locus est nec uberior, quam de officiis, a quibus constanter honesteque vivendi præcepta ducuntur.

Cicero De Officiis, Lib III., Cap. 2.

TO

THE RIGHT HONOURABLE

LORD RANDOLPH HENRY SPENCER CHURCHILL, M.P.,

CHANCELLOR OF THE EXCHEQUER,

AND LEADER OF THE HOUSE OF COMMONS,

This Work

IS

Dedicated,

IN ADMIRATION OF HIS LORDSHIP'S BRILLIANT QUALITIES AS A

STATESMAN, ORATOR, AND LEADER OF MEN,

BY THE AUTHORS.

PREFACE.

In no branch of knowledge has there been of late years more decided progress than in that of Comparative Pathology; and we may venture to say, without fear of contradiction, that upon the still further prospective elaboration of this most important Science, human welfare in large measure depends. To the greater encouragement of original research in the various departments in Pathology by the more enlightened countries, in Europe, in America, and in our colonies, our progress in sanitation and therapeutic knowledge is very greatly due.

The bonds of union between human and veterinary medicine and surgery are yearly—we had almost said daily—becoming more and more intimate, as men are beginning to realise the necessary connection which must subsist between all vital phenomena, whether normal or abnormal. Hence it comes about that investigation in each and every branch of Pathology and Surgery is of the greatest importance, not only in itself, but as bearing upon every other part of each of these two wide subjects.

It was in 1885, that we first put before the veterinary and scientific worlds primarily, and before the public secondarily, "A Manual of the Theory and Practice of Equine Medicine." In the preface to that work, our intention to follow it up by the production of other treatises was intimated. We have it in contemplation before long to issue a work on Comparative Pathology, which is already in hand, and conjointly, "A Manual on the Theory and Practice of Equine Surgery." A complete and comprehensive book on veterinary medicines is already written by Mr. George Gresswell, in co-operation with Mr. Charles Gresswell.

This work, which is now passing through the press, is entitled "The Veterinary Pharmacopœia, Materia Medica, and Therapeutics," and will shortly be published by Messrs. Baillière, Tindall, & Cox.

It will be evident that the large amount of investigation, necessary before writing such works as these, is only to be accurately estimated by those who have devoted their special attention to similar pursuits. In the midst of professional calls, it is a matter of great difficulty to find sufficient leisure—not to speak of the question of remuneration—for the necessary application. For these reasons, our purpose of bringing to completion a work we have had in view, has not yet been accomplished. Recently we have been engaged in the study of the malignant tumours of men and animals, in the hope of shedding some rays of light on the nature and etiology of these insidious and most interesting manifestations of disease; and we hope that our work will, in the future, be not altogether in vain, especially as, working together and separately, we have reason to hope for more complete knowledge, than that we at present possess.

Such marked success as we scarcely hoped for has induced us to continue more quickly than we otherwise should have felt courage for, our deliberately expressed resolve. Of course literary and scientific workers will recognise the great difficulties encountered in working thus rapidly. That we should have been utterly unable to do so, we may with all modesty say, had it not been for the fact that much of what we have given to the world has existed in the form of practical and written knowledge for a considerable period. The treatment recommended in this book, as in the others for which we are responsible, is mainly the result of the prolonged experience of the lifetime of a man who has done very much for the progress of veterinary science. The numerous pupils of the late Mr. D. Gresswell will recognise the painstaking care with which he always strived both to alleviate and to prevent the diseases and disorders of the domesticated animals. The study of science in all its forms was to him the chief joy in life, and he has left what we may with justice call a monument of fame, if only by the impress he has made on the veterinary science of his day.

The great encouragement we have hitherto received from the press, and the large sale of our previous works, have been to us at once most gratifying rewards for labours achieved, and at the same time have furnished us with a most wholesome stimulus for renewed application. The need of and consequent demand for such a handbook as the present one have been abundantly testified in numerous ways.

With regard, however, to the way in which this particular work came to be written, we may say that although we had intended to bring out before long a book of this character, still we should in all probability certainly not have carried out our intention so rapidly, had it not been for the enlightened and most kind courtesy of the Editor and Proprietors of *The Yorkshire Post.* We hope that the readers of that well-known and justly esteemed paper will have derived as much advantage as we ourselves have done. Were it only for the exigencies of providing "*copy*," a most wholesome, if at times a very inconvenient stimulus to a writer, we ought to acknowledge the benefits we have derived from our connection with this widely circulating paper. The advantages, however, are by no means confined to this necessity, and we can only say that we hope the good accruing has been shared by all others as well and as much as by ourselves.

In view of the large amount of errors prevalent regarding the diseases of the horse, and the very great detriment often accruing in consequence, we do not apologise for again intruding on the public, inasmuch as our efforts hitherto have met with the approving commendation of many of those justly entitled to form an opinion as to the merits of our work. When the favourable remarks of the general public, to whom we also appeal, were added, we no longer feared that this, our latest production, would meet with a reception no less favourable than that of its predecessors. The number of letters we have already received from the readers of *The Yorkshire Post*, justify, we think, this anticipation of ours.

It is well to add that most of the cases described have actually occurred in the practice of Mr. J. B. Gresswell, and that the treatment mentioned and recommended is that usually carried out by him. The whole of the literary and pathological portion of the work has fallen to the lot of Dr. Albert Gresswell; but the principal part of the recent revision, alteration, and correction, of the whole work as opposed to the original articles as they appeared in *The Yorkshire Post*, has been carried out in co-operation with Dr. Gresswell, by Mr. George Gresswell, who also has re-written certain portions here and there, both in the original articles, and in the body of the book as it now stands, as well as the introduction.

To Professor J. M'Fadyean, of Edinburgh, we are indebted for the two valuable illustrations of the horse's brain, and to him also we owe our frontispiece and the description of it.

In addition, we have in conclusion, to acknowledge our indebtedness to the following :—

Firstly and chiefly, to the prolonged and extensive experience of the late Mr. D. Gresswell; and also to the admirable and classical researches of Drs. Fleming, Klein, and Cobbold, to Messrs. Percivall, Williams, Gamgee, Signol, Charles Gresswell, Mayhew, Brown, Chauveau, and others.

ALBERT GRESSWELL,
Kelsey House, Louth.

GEORGE GRESSWELL,
Mercer Row, Louth, Lincolnshire.

JAMES BRODIE GRESSWELL,
Veterinary Institute, Louth, Lincolnshire.

July, 1886.

CONTENTS.

——o——

Part I.—Medical Diseases of the Horse.

CHAPTER III.

CHAPTER IV.

CHAPTER V.

CHAPTER VI.

CHAPTER VII.

Part II.—Surgical Disorders of the Horse.

CHAPTER I.

CHAPTER II.

CHAPTER II. *(continued).*

CHAPTER III.

CHAPTER IV.

CHAPTER V.

CHAPTER VI.

CHAPTER VII.

CHAPTER VIII.

CHAPTER IX.

CHAPTER X.

INTRODUCTION.

———o———

In these times of severe depression, depression which is certainly very strongly felt by agriculturists, and those who are connected with agriculture, no subject in the nature of an introduction to a work dealing with the more common diseases of the domesticated animals, is more likely to prove of interest and value, than a plain exposition of some of the erroneous views, which are commonly held with respect to the nature, prevention and curability of certain maladies. The knowledge of the phenomena of disease among those who have not carefully studied them, must necessarily be far behind that of the skilful specialist, who advances with the genius and spirit of the times. Some of the opinions held by the public are a source of much evil ; in many instances these errors have been impressed on the minds of the people by their leaders in past generations, and now the uprooting of them proves a slow and tedious process, which still bars the path of progress.

When the historian of the future takes in hand to record the wondrous discoveries made in this latter half of the nineteenth century, he will have a very pleasant and a very lengthy task ; for they have truly been great and marvellous. In the field of practical science, the development of our knowledge of the marvellous and varied properties of electricity, and the inestimable value of these discoveries in everyday life, will no doubt attract deep attention.

In human and veterinary medicine, the elaboration by Pasteur and others of the germ theory of disease originated by Astier, Schwann, and Cagniard-de-Latour in the first portion of this century, marks one of the most important epochs in the history of these sciences. The influence exerted by these discoveries is immense. Almost equally great is the revolution in the modes of treatment of disease, and this is to be attributed partly to that particular practical contribution to the subject of antiseptic measures for which we are principally indebted to Sir J. Lister.

These points hold equally with regard to animals and to man. It is not too much to say that the light, which is thrown by the two departments of medical enquiry on each other, is daily becoming more and more thoroughly recognised.

The microscope has revealed the nature of the 'poison,' or 'contagion,' or 'virus' of the contagious fevers of man and animals; and it is now known that each of the specific fevers runs a more or less definite course, presenting special peculiarities, by which it may be recognised, in accordance with the characters of a particular kind of 'virus,' which multiplies in a most marvellous and rapid manner. Even so long ago as at the time of the 'Great Plague' of London, the belief was expressed that the pestilence was probably due to some living organisms, which entered the blood, quickly multiplied in it, and passed from one individual to another, through the medium of the air ; or still more certainly, if there should happen to be actual contact between the tissues of individuals already affected, and those of other people. In those days, however, men had not the means necessary for the discovery of the minute living organisms, which give rise to the diseases alluded to. At length, however, it has been demonstrated beyond doubt, that many diseases, such as, for instance, glanders, hydrophobia, anthrax, or splenic fever, tuberculosis, popularly known as consumption, that dread malady of man and beast (which is due to the Bacillus Tuberculosis), and others, are severally connected with, and therefore, in all probability, dependent on the presence of different kinds of vegetable fungi, of microscopic size, in the blood and tissues. Moreover, in the case of those fevers in which special germs have not as yet been satisfactorily demonstrated, there is but little reason to doubt that renewed and more searching investigation will lead to a similar conclusion as to their causation. There is, for instance, reason to suppose that dysentery will eventually be proved to be due to some living vegetable germs ; and, although it is doubtful if cholera is due to a similar cause, it is most probable that the discovery of the virus of this disease is but a question of time.

If we contemplate the fact that such horrible plagues among the higher animals, are caused by the inroads of myriads of certain specific germs, we shall find very much food for earnest reflection. In the first place we must remember, that all the higher organisms, both animal and vegetable, are in reality, composed of innumerable cells, which may to a large extent be considered as separate living units. More than this, we find the normal blood of a healthy animal containing millions of little cell-like creatures, e.g., the red corpuscles which are not so very much like living animals, and the colourless corpuscles, which are exactly like those little creatures called amœbæ. Who can say what is the real significance of the presence of these small organisms in the blood,—for organisms they can be without doubt truly called ? How do we know that they, originally living in the outside world, have not gradually succeeded in taking up their abode in the blood of the higher animals ? This is one way of looking at the question, and it must be confessed that it is an admissible explanation of their presence. After many generations, according to this view, they have come to be essential constituents of the higher animals, and to subserve necessary functions. Of course such speculations are beset with difficulties. With regard, however, to the much more simple germs which give rise to disease, does it not almost look as if certain very low forms of life, happening by some combination of circumstances to be favourably implanted into the bodies of the higher animals, run a certain course of adaptation to their new conditions ? At first, being exposed to a new environment, they multiply rapidly to the very great detriment of their hosts. The latter at length react in such a way to the stimuli set up, and so successfully, that tolerance is established, and the disease is not of very great moment after the lapse of generations, when by the influence of hereditarily transmitted adaptation, the organisms are powerless for mischief. This subject, however, is replete with so many difficulties at present, and the ground is so untrodden, that reluctantly we must leave it, and pass on.

In this connection, a few words will shew the dreadful nature of anthrax, and the supreme importance of looking for remedies both curative and preventive of all diseases, which affect man and animals. No one is of greater interest, than that most disastrous scourge, which goes by the name of anthrax. All animals are liable to attack, including birds and even fishes. No clime is exempt from its ravages. In past times, this disease raged as a malignant and destructive epizootic in man and animals. The 17th and 18th centuries were especially remarkable for devastations made by severe outbreaks. In 1617, the malady was so fatal, that over 60,000 people died around Naples, from eating the flesh of animals which had died from the effects of these insidious inroads of the Bacilli Anthracis.

Anthrax in man is known as woolsorters' disease in this country, and also as the so-called malignant pustule, which is developed as the result of local inoculation, produced by handling the wool of animals which have died from anthrax, or by contact of an absorbed or inflamed surface with a diseased carcase. In Northern Asia, it is known as the Siberian plague. Although it does not frequently affect the horse in this country, anthrax is of common occurrence among the equine tribe as Loodianah disease in Central India. It is well-known in Southern Africa, as the Cape horse sickness ; and also in Australia, where it is called the Cumberland disease ; and in North America and South America. As Texas fever, in the United States, it is of frequent occurrence, and makes severe havoc among the cattle there. According to Toussaint, animals of the value of 20,000,000 francs, die annually of splenic-fever in France.

In certain districts of England it is not unknown, being greatly dreaded at times by owners of stock, and with good reason. Have any therapeutic measures been found which will stay the growth of germs, and thus prevent the inroads of contagious diseases ? It is well here to state emphatically what we shall have occasion to reiterate as we proceed, viz. : that very much more is known scientifically, than is dreamed of by the populace. Everyday our knowledge grows. Mr. D. Gresswell administered sulphite of sodium extensively as a preventive in cases which were exposed to the infection of anthrax, foot-and-mouth disease, and cattle plague, and found that this medicine was of great value in the case of the two former diseases, and also of some value in the third. The value of these measures in the case of the first two diseases has been corroborated by ourselves.

We may here quote a few passages from Finlay Dun's " Veterinary Medicines," in order to show our readers what influence this drug has in preventing the development of disease. " Professor Polli, of Milan, made about 300 experiments with the acid sulphite of sodium, mostly upon dogs, and found that it neutralised, or at any rate diminished, the effects of animal poisons. A striking experiment was made with the muco-purulent discharge from a glandered horse. Forty-five grains were injected into the thigh veins of two strong dogs, one of which for several days previously had received two drachms of sulphite of sodium daily. Both became drowsy and panted, but the one protected by the previous administration of the sulphite, although at first seeming to suffer most from the injection, was in a few hours able to eat, and was next day in tolerable health. The other, however, became more drowsy, and stood with difficulty. By the third day the limb was tender ; by the fourth, mortification set in ; and the animal died on the sixth." The importance of these experiments cannot well be over-estimated.

The fact then appears that in the cases of foot-and-mouth disease and anthrax, animals can in many instances be largely protected against invasion by the influence of certain drugs. It must not be supposed that these measures are always effectual in preventing the onset of these diseases ; but it has been abundantly proved that in many cases at any rate they are of great value.

Talking of the prevention of disease, let us here just discuss very briefly the causes of disease.

Perhaps the most important of all is the general inattention to hygienic laws. In former times, there was great neglect of sanitation ; but now, owing to the preventive measures of better and more careful management, all diseases have become less common. In the case of the horse, those which are doubtless due to the multiplication of germs in the blood and tissues, and we include among these strangles, influenza, glanders, farcy, purpura, horse-pox, and anthrax; as well as many other diseases in all animals, are most probably largely on the decrease in this country. There is now more attention to drainage, and the general laws of hygiene are more carefully attended to, than was the case in earlier times.

However, just as all other progress is marked by more or less rhythmical waves, so also in the case of diseases, periods characterised by outbreaks of exceptional intensity and virulence alternate with seasons marked by epidemics of less extent and diminished severity. At certain times, glanders becomes more prevalent among horses, and afterwards it again makes its appearance more rarely. When a contagious disease breaks out in great severity, or when it occurs with more than ordinary frequency, as a rule, the cause or causes can be found. Some flagrant hygienic fault or omission, or the importation of diseases from abroad, is generally at the root of outbreaks of disease among stock. Still the average number of cases which have occurred annually during the last five or six years, is much less than in past times. This is no doubt largely due to the injunctions ordered by the Contagious Diseases (Animals') Act ; but, did not our regard for cleanliness and hygiene alike progress, such laws would have but a temporary value.

A knowledge of the causes of disease is of primary value to owners of stock. We often hear that so-and-so has had a "run of bad luck." This, in most instances, simply means that the animals are suffering one and all, from the effects of certain neglected causes of disease. When these and the ordinary laws of health are known and attended to, disease on the farms and in the stables may be expected to diminish in the proportion of 60 per cent. at least. Prevention is better than cure ; and, the causes of disease being comparatively few and simple, time and money spent with the view of obviating them, is very well invested. Diseases are too frequently brought about by errrors in dieting, and in the amount of work done, which may be either too great or too small. Age, cold, damp, and wet are often productive of disorders. Poisoning is still not stamped out. Injuries due to accidents or to carelessness frequently bring about severe wounds, and many kinds of lameness. There are also many minor causes also, such as worms, tumours, external and internal parasites. It will be seen at once, that many of these fruitful causes might be obviated. There is no doubt that those which are preventible are diminishing, in direct proportion as the knowledge of hygiene and science increases among the rural and urban populace. Errors in dieting are still common ; mistakes of this nature being especially made in the feeding of heavy draught horses. The serious disorders caused by exposure to damp, cold, and draughts are diminishing. More care is taken with regard to proper modes of ventilation, and unnecessary exposure is avoided. Many horses are still poisoned by the ill-advised administration of medicines by the ignorant ; but it is very probable that, as veterinary science has made such a determined and successful advance in the last few years, wholesale quackery will probably soon be a thing of the past.

Congenital defects might often be obviated by the exercise of greater care in the selection of animals used for breeding. Injuries and many causes of lameness might also often be avoided. Tumours to some extent probably depend on inherited tendency, and therefore care in breeding is highly necessary. Finally, many specific fevers might doubtless be lessened by greater attention to the principles of hygiene. For instance, there is no doubt that anthrax, which, as we have said, is liable to affect almost all animals, depends on bad drainage, at least to the extent that animals which are exposed to the effects of insufficient and faulty drainage, are more likely to go down with this disease than animals more favourably situated. Similarly too, with sheep-rot, and other diseases of sheep, sometimes a whole flock of sheep will contract a serious affection of the lungs. When this happens, it is often the case that they have been exposed in bad weather to the noxious vapours of badly-drained lands, reeking with decaying vegetable matter. What else could be expected ?

We have mentioned some aspects of recent progress. Let us turn now to another, which is also of the greatest interest and value.

In the field of Philosophy, the firm establishment of the Evolution Theory as a fundamental basis of thought, is of the highest importance. It is only in comparatively recent times, that the Law of Causation, which had been already for some time more or less completely recognised in the less complex of the phenomena around us, was also applied, principally as a result of the brilliant researches of Mr. Darwin, to the involved processes of living organisms. More recently still, we have been attempting to unravel the varied and mysterious abnormal processes, which are at times exhibited by animals and plants, in accordance with this same Universal Law.

As an explanation of the mode of development of animal and vegetable life, and especially of the former, whether in the case of the individual or in that of the tribe, of the origin of man, and of his language and social customs, the hypothesis of evolution has been of incalculable benefit in the past ; and it is also of the greatest possible advantage in moulding our methods of research to-day.

Marvellously important however as this belief is, it must be emphatically stated that evolution can only be regarded as a method of procedure. Mr. Herbert Spencer insists that phenomena indicate, or are the expressions of an 'Infinite and Eternal Energy'; and it is clear that the result of the deepest reflection can only be to carry us much further in the direction of positive belief. That gradual but definite progress of living organisms from simpler to more perfect forms, manifestly points to some Power who ordains this state of things, this wonderful co-ordination of the intricate processes going on around us. Indeed it may be said, that those who attempt to remove in thought the controlling agency, the guiding Power, cannot refrain from admitting the efficiency of Blind Chance to take the helm. It is obviously quite impossible, completely to explain the causation of phenomena, however simple. The initiation and maturation of new organs and new structures, the phenomena of bodily and psychical development, all indicate the existence of something far higher and greater than we can conceive of, some grand reality of which we only see the superficial manifestations.

It is important to remember that when we have assumed the complete idea of connected and continuous causation, which is called the Evolution Theory, we have by no means eliminated the necessity of belief in a Great First Cause.

It is one thing to recognise a necessary and inevitable connection of sequence between those simple forms of matter, the nebulæ on the one hand, and the most complex forms of material seen in other portions of the universe on the other ; and it is another to believe that herein has been reached a fully satisfactory solution of all things and all mysteries. All that man can ever claim to have done, is that he has substituted the idea of one great Power, unknown and unknowable, for the innumerable spirits and influences which the savage supposed to be the causes of the occurrence of phenomena. Instead of looking upon the great mysteries of life, the unknown factors which have resulted in our presence here as innumerable and indefinite, we now consider them all as expressions of a Great Reality, which we cannot fathom.

Of this ultimate conception, which all men frame for themselves, consciously or unconsciously, it is impossible to rid ourselves, try as we will. In all creeds and in all beliefs, this idea, more or less pronounced, is present:

be they the most anthropomorphic, or be they the most abstruse of religious or scientific dogmas. Even those who regard the Evolution Theory as final—and there are very few, if any, who do so—are logically compelled, as we have said, to introduce an unknown power, for they have recourse to the most unsatisfactory of all factors—that called Chance. Either Chance means practically nothing at all, either it means that facts, which are manifestly great, are to be explained by a very small and insignificant cause, or it really implies a Great Unknown Power. In fact, this belief in one power of some kind, more or less powerful, may be said to be one, in possessing which, all men agree. Such an assumption, is in reality, one of the indispensable conditions of thought. In Evolution we have the dim notion of a method, and, when this is realized, the question still presents itself,—What is it, or rather, Who is it—who guides this process? What causes this gradual growth which is palpably going on, and so far as we can determine, in the direction of Advance and Progress? What causes Evolution? Some *may* be satisfied with the answer, "We don't know." Let us, however, while acknowledging this in some sense, as, indeed, we are compelled to do, look for and recognise a higher ideal, and boldly acknowledge the Almighty and Inscrutable Power which, try as they will, men cannot refrain from postulating in some form at least.

The Evolution hypothesis has exerted the very greatest influence on all the sciences; but on no kind of knowledge has it had more effect than on the veterinary branch. The development of all higher animals from lower types has now been abundantly attested by geological investigations. In common with other animals, the horse, ox, sheep, pig, and dog have arisen from simpler and more highly generalized forms. The gradual production of the horse from creatures having five perfect toes on each limb, has been established, beyond the possibility of doubt. The gradual loss of all toes except the central one, which is now provided with two rudimentary appendages, called splint bones, is one of the facts which show us how the horse has been steadily modified in a definite direction, since the time of what is technically called the Eocene period, until at length this animal has assumed the present well-known shape and proportions. Such facts as these are well known. Great, however, as has been the influence of new ideas, the advantages already derived are immeasurably enhanced by their importance as guides to the methods of modern research.

At the present time, it may be truly said that a wave of knowledge is sweeping many erroneous notions away, though this is not being effected without trouble and some annoyance. Work is not done, and cannot be done, it seems, without a great amount of friction. There has always been, for instance, a certain amount of jealousy between the so-called practical and the so-called theoretical people. It is high time this was done away with. It is a common belief among the populace, and even among some of the most highly educated—and there is no class of men who adhere to this delusion more rigidly and obstinately than many of those who have to do with horses—that there are two distinct divisions into which all knowledge may be divided, viz.:—theory and practice. Perhaps no opinion has stayed the progress of veterinary science among the general public more than this one. The tendency which besets the earnest student to confine his attention to recorded facts, rather than to the practical observation of them, is one of the most serious evils, and one which we should always be on guard against in every way ; but, although we might inveigh with emphasis against the fearful cramming which goes on at the present day, constituting one of the worst features of modern progress, we must, with still more animation, point out that there are immense storehouses of knowledge for the busy student, and immense fields of research for the industrious, in almost every department of science. By years of painstaking care and labour, men like Pasteur and Klein, and many others, are opening up vistas of new worlds of knowledge. We are just beginning to peep through the dim apertures in the wall of ignorance, and catch glimpses of the truth. This is the kind of knowledge, and this is the kind of work which the ignorant will often condemn as theoretical.

On the other hand, the so-called practical man, whose actions from day to day perpetuate the grossest ignorance and the worst delusions, is often extolled. Let us not be misunderstood. This self-styled practical horseman in many instances is not practical at all. His vaunted practical ability, being based on false theory, is worse than useless. Perhaps, if he is a horseman, he is one who will buy a horse with the most palpable defects ; perchance he will fail to recognise the symptoms connected with a diseased spinal cord, or he may possibly purchase a roarer, or a horse lame in both fore feet, and come home thinking he has made a good bargain. He will perhaps tell you that intestinal worms are rather advantageous, than productive of injury ; and that some diseases, such as strangles, should not be *interfered* with. According to such a one, the trainer's knowledge is more useful and reliable than that of the cultured specialist, who has spent years of research into the actions of the organs and tissues in health and disease, into the value of the various remedial measures, by which abnormal processes can be controlled. In short, it has been advanced that the amateur is as highly qualified as the skilled professional man.

It is thought by some, who forget or do not know the intricacy and complexity of diseases and their varied characters, that a prescription which has been given in special circumstances, to a special case, shewing particular characters, can be freely used again by the owner or even by the groom, if only he imagines there is a similarity, and that it might be useful. No delusion is more strange than that which induces some to act habitually as their own veterinary surgeons. The folly of letting animals die from want of proper attention, is as extreme as that which prompts the owner to undertake the doctoring of his own stock. It would seem unnecessary to state that the strangely-involved symptoms of disease cannot possibly be understood by an unpractised observer, if it were not a fact that great annual losses are involved by want of proper care and attention. If the veterinary surgeon was recognised now, as he ought to be, and will be in times not far distant, a man would as soon think of making his own boots and blacking them himself, and cutting his own cloth, as he would of doctoring either himself, or his horse, or his dog, or his cow, when *seriously* diseased, or of allowing his man-servant to do so. Moreover, the test of experience will show that he who selects the best procurable professional advice will be the pound-wise man, while the other will be the penny-wise one. Sometimes, of course, it may be found impracticable to procure the services of a good practitioner; and then a useful handbook is almost indispensible to the large owner of stock. It is needless to add that all the general conditions of hygiene and management should be well understood by all those who undertake the care of animals of whatever kind.

Let it not be thought that we ignore or undervalue the only real knowledge, that gained by the actual and practical observation of facts. So far from doing so, we would most earnestly recommend that every man should mainly aim at the acquisition of precise and real scientific culture. There is, however, a very great distinction to be drawn between true practical power, and that which often among the populace passes as such. No knowledge can be complete which ignores many of the conditions of a problem. In the case of living beings, a lifetime's education is not too much to be spent in the acquisition of the knowledge of their extraordinarily varied phenomena in health and disease. The nervous, muscular, vascular, circulatory, respiratory, digestive and sketetal mechanisms, are each most elaborately constructed. Scientists are just beginning to understand a l these things. The amount of knowledge however which we possess, grows daily. Every year brings forth its startling discoveries of the phenomena presented by living things. Every year, new drugs and new curative and preventive appliances are being discovered. Every year alters to some extent our views of disease. The man who is now marching with the times, may find himself in two or three years sadly lagging behind in the rear, and yet much of this knowledge is condemned by the unskilled as theoretical. Could anything be more palpably absurd?

In this connection, it may be remarked that the veterinary profession is not always without blame. The value of the scientific and theoretical knowledge, which must be gained at one of the veterinary colleges, it would be impossible to over-estimate. At the same time it must be remembered that the number of patients at any given college is not very large. A student can, therefore, scarcely be said to have completed his education, unless he has had a considerable amount of actual practice. This may be done, and is very generally effected, before entering at a college, by becoming pupil to a veterinary surgeon in large practice, for at least a year.

We have spoken of some popular errors. Now let us also consider some ancient customs of treating disease in animals. The peculiar and superstitious notions still prevalent, chiefly among the rural populace, regarding disease and its treatment are very marvellous. Some have a very ancient origin, while others are of more recent growth. Many are grotesque in the extreme, ·and some very harmful. The erroneous views of the public on veterinary matters are a source of much harm. Error is pregnant with evil fruit. Now, however, the fascinating sway, long held by superstition, is slowly but surely retreating, like the morning clouds, dispersed by the rising sun, behind the firm onward march of education.

The actual origin of many of the erroneous views still held by some, it is impossible to ascertain ; but in order to illustrate the antiquity of many widely known myths, which have actually forced their way into history, and passed without question for a long time, just as bad coins will now and then pass among good ones, we may quote a few passages from Edward Clodd's "Childhood of Religion." First, with regard to the ancient myth of William Tell, this author writes—"The story is well known how in the 1307th year after Christ, the cruel Gessler set a hat upon a pole as a symbol of the ruling power, and ordered everyone who passed by to bow before it ; a mountaineer, named Tell, refused to obey the order, and was at once brought before Gessler. As Tell was known to be an expert archer, he was sentenced by way of punishment to shoot an apple off the head of his own son. The apple was placed on the boy's head, and the father bent his bow. The arrow sped, and went through the apple. Gessler saw that Tell, before shooting, had stuck another bow in his belt, and asked the reason. Tell replied ; 'To shoot you, tyrant, had I slain my child.' Now, although the crossbow, which Tell is said to have used, is shown at Zurich, the event never took place. One poor man was condemned to be burnt alive for daring to question the story ; but the poor man was right. The story is told not only in Iceland, Denmark, Norway, Finland, Russia, Persia, and perhaps India, but it is common to the Turks and Mongolians, 'while a legend of the wild Samojedes, who never heard of Tell, or saw a book in their lives, relates it, chapter and verse, of one of their marksmen.' In its English form, it occurs in the ballad 'William of Cloudeslee.' The bold archer says:—

> I have a sonne seven years old,
> He is to me full deere ;
> I will tye him to a stake—
> All shall see him that bee here--
> And lay an apple upon his head,
> And goe six paces him froe,
> And I myself with a broad arrowe
> Shall cleave the apple in towe.

The story is an old Aryan sun-myth. Tell is the sun-god whose arrows (light rays), never miss their mark, and likewise kill their foes. There is another old tale, over which I have cried as a boy. You have heard how the faithful dog, Gellert, killed the wolf which had come to destroy Llewellyn's child, and how, when the prince came home, and found the cradle empty, and the dog's mouth smeared with blood, he quickly slew the brave creature, and then found the child safe, and the wolf dead beside it. At Beddgelert, in North Wales, you may see the dog's grave neatly railed round.

"Now this story occurs in all sorts of forms in the folklore of nearly every Aryan people, and is found in China and Egypt. In India, a black snake takes the place of the wolf, and the ichneumon that of the dog ; while in Egypt the story says that a cook nearly killed a Wali, for smashing a pot full of herbs, and then discovered that among the herbs their lurked a poisonous snake. It is safe to conclude that marvellous things, which are said to have happened in so many places, never happened anywhere."

When we consider the origin and gradual evolution of stories such as these, and the way in which they are handed down from father to son through innumerable generations, we can hardly be surprised that certain notions of disease and its treatment have become so ingrained in the minds of many, that time and education alone will be able to efface them.

We may here give a few examples. According to the rural populace of many parts of England, there is a disease of cattle, called the tail worm, also spoken of by some as the "wolf." It is believed to be discovered by a softness between some of the joints of the tail. In these cases it is believed that it is necessary to slit open the under surface of the tail, and to rub in a certain mixture composed of salt, wood-soot, and garlic. When these absurd notions arose, one cannot tell, but although it has long been known to veterinarians that there is absolutely no such disease, yet the practices here described are still largely carried out. Moreover, many well-educated farmers and others also believe in the actual existence of the tail worm, and in the necessity for these ill-devised practices. That many practices such as these had their origin in superstition is certain. The old practice, now abandoned, of placing a live frog down the throat of a cow, for the cure of a certain malady, was commonly adopted at the commencement of the present century ; and is put down as a recognised plan of treatment in the Compendium of Farriery, published in 1790. The only wonder, in looking through old and erroneous notions, is why certain of them are in full force *to-day*, while others are as entirely given up.

The evolution of the well-known story of Cinderella is so interesting, that we cannot help again quoting some passages from the above writer relating its origin in full. This author writes :—"Let us see whether Cinderella is a British-born lady in disguise, or whether she came from some very old nursery in the East. She must have come therefrom, for we find the framework of the story in the Veda, where Cinderella is a *dawn maiden!* The aurora in her flight leaves no footsteps behind her, but the prince, Mitra—one of the Vedic names for the sun—while following the beautiful young girl, finds a slipper which shows her footstep and the size of her foot, so small that no other woman has a foot like it. This sun myth, which tells of a lost slipper, and of a prince who tries to find the foot to which it belongs, and who cannot overtake the chariot in which the maiden rides, is the source of the dear old tale. Cinderella, as you will remember, was beautiful only when in the ball-room or near the shining light. This means that the aurora is bright only when the sun is near ; when he is away her dress is of sombre hue—she is a *Cinderella.* The Greek form of the tale says, that whilst Rhodope was bathing, an eagle snatched one of her slippers from her maid, and carried it to the King of Egypt, as he sat on his judgment seat at Memphis. The king fell in love with the foot to which the slipper belonged, and gave orders that its owner should be searched for, and when Rhodope was found, the King married her.

"In the Hindu tale a rajah has an only daughter, who was born with a golden necklace which contained her soul, and the father was warned that, if the necklace were taken off and worn by another, the princess would die. One birthday he gave her a pair of golden and jewelled slippers, which she wore whenever she went out ; and one day, as she was picking flowers upon a mountain, a slipper came off, and fell down the steep side into the forest below. It was searched for in vain ; but not long after a prince who was hunting, found it, and took it to his mother, who judging how fair and high-born the owner must be, advised him to seek for her, and make her his wife. He made public the finding of the slipper throughout the kingdom, but no one claimed it, and he had well nigh despaired, when some travellers from the rajah's country heard that the missing slipper was in the hands of the prince, to whom they made known its owner's name. He straightway repaired to the rajah's palace, and showing him the slipper, asked for the hand of the princess, who became his wife. After her marriage, a jealous woman stole the necklace while she was sleeping, and to her husband's deep grief her body was carried to the tomb. But it did not decay, nor did the bloom of life leave her sweet face, so that the prince was glad to visit her tomb ; and one day the secret whereby her soul could be restored was revealed to him. He recovered the necklace, placed it round her neck, and with joy brought her back to his palace. The like framework of a slipper for whose pretty wearer a search is made, and who becomes the finder's wife, occurs in the Serbian tale of 'Papalluga;' in the German tale of 'Aschen-puttel;' in the fable of La Fontaine about the 'Milkmaid and her Pail;' and other varients of the story, whose birthplace, as we have seen, was in Central Asia."

In looking through the various superstitions of different races, one expects to come across a number regarding horses, cattle, frogs, and toads. In the *Nineteenth Century* for July, the writer on the article on Transylvanian Superstition tells us that a toad taking up its residence in a cow byre, is assuredly regarded as in the service of a witch, and has been sent there to purloin the milk ; and that it is necessary, therefore, that it be stoned to death. The skull of a horse also placed over the gate of the courtyard, or the bones of fallen animals buried under the doorstep, are preservatives against ghosts. In our article in *The Yorkshire Weekly Post*, of August 22, we spoke of the absurd practice of burning the palate of the horse for the supposed cure of lampas. This cruel method of treatment, though gradually dying out, is still not very uncommon, being generally carried out by the village blacksmith.

We might give many other examples of erroneous practices which are in vogue, but we have already diverged somewhat from our original point, and those mentioned shew us that education is alone able to dispel illusion. Quackery of many forms and varieties is founded on superstition, and a belief in the efficacy of many absurd modes of treatment and ill-compounded mixtures is nothing more than superstition. Thanks to the efforts of the Royal Society for the Prevention of Cruelty to Animals, many of the more cruel methods of erroneous treatment are being put a stop to. In Japan, and certain other countries, every patentee of an advertised medicine is required to submit an analysis of its contents to the Government. If the mixture is deleterious, he is not allowed to dispose of it ; whereas, if good, he may proceed with his business. The Japanese have thus to a large extent emancipated themselves from the thraldom of quackery. Scientific education is daily becoming more appreciated, more honoured, more revered throughout the world ; and as it spreads, quackery, superstition, and all unfounded beliefs, must eventually sink into oblivion.

Perhaps one of the worst of all errors, are those in accordance with which it is dogmatically stated, that certain curable diseases are incurable, and certain preventible maladies non-preventible.

Many of our readers will doubtless be aware that beasts are subject to a certain disease of the tongue. The true nature of this affection, which is characterised by the growth of tumours of varying size on this organ, has only recently been determined. The affected animals slaver profusely, and lose flesh rapidly. As the tumours grow, the breath becomes fetid, the emaciation becomes still greater, and the animal is unable to take any solid food. This disease, termed actinomycosis, because it is caused by the growth of a fungus called actinomyces in the tongue, is popularly supposed to be incurable. This is a very great mistake. The malady is in almost all cases curable by judicious treatment. In order to explain the symptoms and treatment of this emaciating disease to our readers, we may quote our remarks on this subject, recorded by us in Dr. Fleming's *Veterinary Journal*, and in the *Veterinarian*. A full exposition will doubtless be of interest to most of our readers, for, although the disease is not a very common one, it is capable of spreading among a whole herd of cattle, in a comparatively short space of time. Moreover, it will serve to illustrate how unfortunate may be the result of the notions of those, who take for granted that curable diseases are not amenable to treatment. The first case to which we wish to call attention is one of a two-and-a-half year old red bullock, the property of a cattle dealer, who had kept and fed him on the Lincolnshire Wolds. It was first noticed that the beast, which had fed badly for some time previously, was slavering profusely. He would eagerly champ and chew his hay and seeds, and would then throw them out of his mouth again. When the animal had been ailing for about three weeks, we were called in. The tongue at the time was so bad, that the animal could eat no solid food. At the same time there was a heifer in a similar condition, and two other bullocks were also slightly affected. All had several hardened, yellow, nodulated masses on their tongues. In the animal to which we were called, the tongue was much enlarged ; and was very tender to the touch, and the sides and back of it were studded with nodules varying in size from a marble to a pigeon's egg. One at the back of the mouth in particular was very large, with a superficial erosion. The animal lived solely on mashes and linseed gruel, and at this time weighed about forty stones or under, in the view of the owner. The heifer was killed; but it was decided to adopt curative measures in the case of the bullock. We ought to mention that the breath was fetid in these cases, especially in that of the bullock, as it very commonly is in severe cases of actinomycosis. The treatment ordered at first consisted of painting over the affected part with tincture of iodine, and the internal administration of tonics. The animal, however, made no progress towards recovery, and on March 12th, had shrunk so much, that it was deemed advisable to have him killed. It was, however, eventually decided to continue treatment; and, accordingly, the animal was cast and the tongue was carefully examined. Into each nodule an incision was made, and the cut surfaces were painted over with a mixture of carbolic acid and iodine (iodised phenol). Some of the nodules were seen to be of a light yellow colour, whilst others were of a deeper orange. On the 16th of March, the beast was again seen, and on examining the tongue it was found that all the incisions were healed. In many places the nodules were smaller, and some had quite disappeared. On March 21st, we saw the bullock for the last time. He was very much better, and could eat hay and straw. In April, he was turned out to grass. On September 14th, the dealer saw the animal, and reported him as quite well, and calculated his weight at not less than seventy stones (an increase of thirty stones).

During the past twelve months we have seen a great number of such cases, some of a mild description, others very severe; and, with the exception of two which we ordered to be killed, they have all recovered. These cases, therefore, when thoroughly taken in hand, almost invariably recover.

Again tetanus or lockjaw is regarded by many as a nervous affection. We, however, shall treat of it as a general constitutional disease. The horse is the most susceptible of all the domesticated animals to attacks of this dreadful malady. Tetanus, undoubtedly a very severe disease, is erroneously supposed by many to be incurable. The mortality, however, does not exceed sixty per cent., and in cases not following injury, our statistics show a still less rate of mortality.

Just a few words on the subject of influenza, will not be disconnected with the subject. There are two popular errors regarding this common disease. One is that horses which suffer from a mild attack can be worked with safety during the progress of the affection. The other is, that quack medicines can be profitably used, and by the unskilled, with a view to its cure. There is, perhaps, no malady so well known by those who have to do with horses, and none so commonly met with, as that which goes under this name; it is a disorder which varies considerably, both in character and in intensity. Sometimes it occurs in a very mild form, and sometimes it is of a very severe type. As a rule, influenza is of a more marked kind in large towns, than in country districts. It may be said that it far more commonly leads to a fatal result, than the knowledge we possess of the proper treatment would lead us to expect. This is merely because in many cases professional advice is not resorted to at a sufficiently early stage. In some districts, it is not uncommon to work cart-horses throughout the course of the disease, when its mild nature (seemingly) allows. Of course this is extremely foolish and cruel conduct, and now and again a horse is utterly unable to endure such barbarous measures. The symptoms become more extreme ; and, unless the animal is most carefully tended by the experienced veterinarian, he will succumb. Until the active and acute symptoms have abated, rest is one of the primary essentials in this, as in nearly every disease. Again, of all the maladies to which the horse is subject, influenza is one of the most weakening and debilitating; and yet, ignorant people will still pour down the throats of their long-suffering animals, the most abominable quack mixtures, which very commonly lay the foundation of a fatal result. These mixtures frequently contain large doses of lowering or sedative drugs, which the already enfeebled system cannot resist. Tincture of aconite is but one of the deleterious ingredients of some of the medicines, which have frequently well-nigh brought the poor sufferers, to whom it has been administered, to death's door. It is not uncommon to find animals poisoned by excessive doses of sedative, and other drugs.

We may incidentally remark, that the great fact to learn about influenza, is, that good support in the shape of oil-cake, gruel, and other nutritious food, together with tonics and stimulants, are requisite at a very early stage in the disease. In that debilitating form of influenza, known as pink-eye, some form of alcoholic stimulant is of paramount importance throughout the disease.

A gentlemen once purchased a horse, as he thought, very cheaply at a fair. When the bargain was made, the purchaser inquired if the animal had any defects, for which he was disposed of at so low a rate. The seller replied that the horse had but two faults, the first being that when out at grass he could not be caught. As, however, it was not intended to turn the horse out to grass, this was not a fault in the opinion of the new comer, who at this juncture excitedly asked, " But, tell me, what is his other defect?" " It is this," replied the dealer, " that when the brute is caught, he is no good to anyone." But too often men pay cheaply, as our friend did for his horse, for medicine composed they know not how, which is often useless, and sometimes worse than useless. Recently we were called in to two teams of cart-horses, each one of which had received a ball. All the animals were nauseated, and were not fit for work for several days. This is a very mild instance of the ill-advised treatment, which is too frequently adopted.

We cannot too emphatically condemn the practice which is too general among horse owners and others, of giving and applying medicines and medicaments without professional advice. There are, however, exceptions to all rules, and there are of course some, though very few patent remedies, which are really of use. We are so often asked our opinions concerning some of them, that we think it well to append here formulæ, which will be found of great value for general use. One ounce of methylated spirit, and one ounce of Goulard's extract of lead, make with the addition of eight ounces of water a valuable cooling lotion. This will be found useful for sprains and bruises, where there is no external wound. Another good cooling lotion may be made of one ounce of chloride of ammonium, and one ounce of nitrate of potassium, dissolved in one pint of water.

Regarding liniments and embrocations, the ordinary soap liniment of the British pharmacopœia is very useful as a mild counter-irritant. The compound linament of camphor is also very useful ; it is stronger and more expensive. A good liniment for general use is made up of strong solution of ammonia four drachms, methylated spirit one ounce and a half, oil of turpentine six drachms, soft soap one ounce, hot water sixteen ounces. We must warn our readers against using those embrocations, whose property is to heal and cure all forms of disease to which the horse is liable. They are frequently expensive, and often too strong for general use. Consisting sometimes largely of turpentine, they often have a very deleterious effect. We were called a short time ago to a horse, valued at £80, which had sustained a slight injury of one of the fore legs. The owner had rubbed in some strong embrocation, and had thus set up acute inflammation. Acute erysipelas set in rapidly, in spite of all that could be done, and the animal died in three days. Never employ an embrocation to a raw wound of recent standing.

Regarding blisters, the ones most generally useful are the red ointment of biniodide of mercury (made of red iodide of mercury in fine powder one ounce, and olive oil one ounce ; mix thoroughly with a wooden knife, and add of melted lard seven ounces) ; and the cantharides ointment (made of powdered cantharides one part, venice turpentine one part, resin one part, lard four parts—melt together). When a mild liquid absorbent blister is required, the liniment of iodine is useful in reducing glandular swellings.

We are often asked to give our opinion of the value of certain liquid preparations for the cure of splints, side-bones, ring-bones, and spavins. No doubt they are of some efficacy for the purposes for which they are used ; but similar and still more valuable mixtures can be obtained at a very much less expenditure. It is the old tale of quack medicines, which people think good because they are dear, or are well spoken of. We append a formula, which can be made up at any chemist's, which is a still more effectual mixture than those we are speaking of. Some may prefer to waste their money, but we trust our readers will not uselessly throw theirs away. Take forty-five grains of perchloride of mercury, dissolve in two ounces of methylated spirit, add forty grains of biniodide of mercury. This (which must be labelled poison) is a very effectual mixture for the reduction of all splints, or other bony enlargements.

We are frequently asked to express our opinion of the value of certain internal patent medicines for horses. We can only speak favourably of two, which we employ ourselves. One is Dr. Collis Brown's chlorodyne, the other is Dr. Blumendorf's worm specific, a very valuable medicine for expelling worms in horses and dogs. It is not cheap, but is very effectual.

To return to our original point, we may say in conclusion, that it is a grievous popular error to believe in quack mixtures, which are said to be universal specifics for all diseases. Let us turn to another fallacy. Our readers may have read Mr. Mayhew's just denunciation of the absurd system of slanting pavements ; but, for the sake of those who do not happen to have seen his remarks on this subject, we may quote a few paragraphs from his work on " Horse Management."

" Some sad and patient animal on a slanting pavement, may have been silently watching, longing for the absence of the groom during a considerable period. No sooner does the creature hear the door slam, than he begins to take small steps backward. The horse thus feels its way, till the sudden fall on the pavement announces that the posterior hoofs have reached the gutter, within the hollow of which the toes are immediately depressed. Such an attitude being attained, all stress upon the flexor tendons is removed from the hind legs. The bones, while the toes can be depressed, sustain the weight of the haunches. Partial ease is thereby received, and with the new sensation, a numbing torpor creeps over the animal. Its feelings are soothed by present pleasure, and the nerves thrown off their guard, grow dead to all outward impressions. The victim of former ages, when taken from the rack, must still have endured agony ; but the lull occasioned by the cessation of acute torture, threw the sufferer into a lethargy, which is reported to have resembled the luxury of a sleep. So it is with the horse. The fore feet are still undergoing torment ; but, under partial relief, the animal seems to doze, or become unconscious to external agencies. The horse, however, has not only to stand, during the day-time, upon a slanting pavement, but it must throughout the night be in this position. Did the reader ever attempt to repose upon a bed slightly out of the horizontal ? The sensation communicated is an incessant fear of slipping off. The sleeper is constantly wakened up with a vivid impression that he is falling, or has fallen on to the floor. The night is passed in discomfort. What is the excitability of a human being, when compared with the fear which haunts the most timid of all timid lives ? Assuredly he should have possessed an enlarged capacity for evil, who first conceived the notion of making a living creature, conspicuous for its strength, its activity, and its timidity, exist in a niche, have its head tied up by day and by night, and be subsequently doomed to rest on a floor, sloping in a painful and unnatural direction."

Rest is of the greatest importance to all living things, and especially in disease, is of the highest utility. Hence, anything which interferes with a horse's, or any other animal's repose, is most strongly to be reprehended. During rest, the waste undergone by the various tissues and organs of the body is repaired. The products of work are removed, their place is taken by fresh material, ready to do work, and the various parts of the body are thereby restored to a condition suitable for the performance of their functions. Diminution of rest, like overwork, will gradually bring on disease. The loss of a night's rest will unfit an animal for the next day's work. The horse sleeps but little, probably not more than five or six hours in the twenty-four, does not always sleep, nor even invariably lie down when taking rest. Some horses which are apparently able to recuperate their energies in a standing posture do not lie down.

In some cases it may be that they have stiff backs, and hence experience a difficulty in rising. Others, having once been halter-cast, will not assume the recumbent position, so long as their heads are tied.

Although a horse may for a long time stand still, the recumbent position is certainly the one in which most repose is obtained ; and a horse never works so easily, nor wears so well, as when he spends six or seven hours daily in this way. The legs and the joints are, under these circumstances, not so liable to become stiffened. Some horses, which will not in the general way lie down in their stalls, may be induced to do so by transferring them to a comfortable, loose, and well-bedded box.

With regard to rest, just a little more may be profitably said. Too much rest is as damaging as too little.

A writer in the *Graphic*, of August 29th, 1885, speaking of " summering " hunters, gives some good hints. He writes regarding these points : "Already we hear the preparatory notes of the next fox-hunting season, and within a few days in more than one district cub-hunting will begin, in order to scatter the litters, and teach the newly entered hounds somewhat of their regular business. All horses, too, which have been 'summered' after the olden fashion, by being turned out to grass for some months, must now be taken up, and gradually got into condition to give any hope of their being at all fit for their work by the winter. Happily, better counsels now generally prevail in equine management, than those which were considered orthodox but a few years ago. The turning out of hunters in meadows to be tormented by flies, or shutting them up in out-houses or large barns, to lead wretched, monotonous lives, and perhaps contract diseases brought on by inactivity, are now practised at a discount." This is very true, and we may say with the above writer, that the best plan is always to give regular, but not exacting exercise.

We have now completed our introduction, and we hope that this short sketch of the present and the future, as compared with the past, may do something to dispel some of those illusions which, like dark clouds, still remain to stay the advancement of the Art and Science of Veterinary Medicine and Surgery.

PART I.

Medical Diseases of the Horse.

CHAPTER I.

GENERAL DISEASES OF THE HORSE.

Influenza. Strangles. Glanders and Farcy. Anthrax or Charbon; Anthracoid Diseases, Glossanthrax and Anthracoid Angina. Scarlet Fever and Purpura. General diseases of the Horse due to errors in dieting and management: Weed or Lymphangitis; Diabetes Insipidus, Diabetes Mellitus; Oxaluria; Azoturia. Tetanus or Lock-jaw. Rheumatism, Acute, Chronic, and Muscular. Rabies or Hydrophobia.

INFLUENZA OR DISTEMPER.

THERE are perhaps few diseases to which the horse is liable so well known as "influenza" or "distemper," and there are few which vary so widely in intensity and diversity of form as does this protean malady.

This disease was first termed "influenza" in the seventeenth century by the Italians, who attributed its origin to the influence of the stars. In addition to the more popular names, influenza has also been called "the epidemic" and "epizootic catarrhal fever," and has also received special appellations, such as "pink eye" or pneumo-enteric fever, according to the form and character which it assumes. Even so early as the beginning of the fourteenth century, influenza is recorded to have broken out at Seville, and many horses then fell victims to its ravages. In the years 1688 and 1693, severe epidemics of influenza occurred throughout Europe, and they were followed by the appearance of a similar fever in man.

Influenza is widely distributed over the Old and New Worlds, and when once manifested shows a marked tendency to travel, usually westward. The parts more particularly affected by the disease are the lining membranes of the nose, throat, and upper part of the wind-pipe, and its continuation into the lungs.

Of the actual causes of influenza so little is definitely known that we shall not perplex our readers by promulgating uncertain theories. As far as we know, however, influenza has not yet been proved to be influenced by atmospheric or astronomical causes ; nor is its origin in any way connected with the special geological conformation of the strata of the earth's surface.

It is nevertheless usually more prevalent at spring time and autumn than at other seasons, though it may appear at any time of the year, and already during the month of February, 1886, the writer has had under treatment many horses attacked by a very severe though not fatal form of the disease.

There is no doubt that defective sanitary arrangements predispose the horse to attacks of this disease, or render the system a more fit receptacle for the growth of the germs of influenza, and no doubt also enhance the severity of the symptoms which are manifested. We would specially draw attention to insufficient ventilation and bad drainage. A due supply of pure air is of paramount importance to the well-being of the horse, while on the other hand small, dark, stuffy, badly-drained stables predispose him to all forms of disease. It has been observed that horses are, as a rule, affected with a more severe form of influenza in large towns than in country districts, and this is doubtless largely due to the smaller size of the stables and the bad ventilation. Indeed, defective drainage and ventilation are sufficient of themselves to produce fatal disease, by causing the animal to breathe air contaminated by poisonous effluvia and emanations. Furthermore, neglect of any description, as well as insufficient food and excessive work, predispose the horse to severe attacks of influenza.

Many of our readers are doubtless acquainted with the more usual symptoms of influenza. The dry staring coat, the coldness of the extremities, and the redness of the lining membranes of the nose, are early manifestations. The temperature, which should reach but 100.5 F. in health, is raised three degrees or more ; and the number of the beats of the pulse, which should number but 36 in a healthy horse, rises to 50 or 60 beats or more per minute. There is sneezing and hacking cough, and from the nostrils there runs at first a thin glairy fluid, which, as the disease is established in two or three days, becomes thicker and more abundant. The temperature often rises a degree or two higher, and the cough also becomes more severe and distressing, while the pulse is still quicker and more feeble.

The breathing of the animal is also increased, the appetite is impaired, and sometimes almost totally lost. Soreness and swelling of the throat cause pain and difficulty in swallowing, and the excreta become more scanty. After about three to six days the symptoms usually abate, leaving the horse much enfeebled, but the strength usually returns in about a fortnight from the first onset of the attack.

Influenza does not by any means always thus speedily terminate in recovery, for in some instances "bronchitis," or inflammation of the bronchial tubes, which are the continuations of the windpipe, may supervene, more especially in weakly and debilitated animals. In these cases the danger is seriously increased—the breathing becomes very difficult, the nostrils are widely opened, and the lining membrane of the nose becomes of a livid purplish hue, owing to the fact that the blood is no longer properly aërated in the lungs. Such cases as these are attended with great risk, and may prove fatal in a week or so. They require all the care of the scientific veterinary surgeon, and their treatment cannot be undertaken by amateurs. Again, influenza may attack the lungs themselves, causing a dull soft cough, and great prostration and acceleration of the pulse ; and sometimes, though rarely, the disease may terminate in mortification of the lung tissue itself. In some epidemics, "pleurisy,' or inflammation of the lining membrane of the chest and lungs, complicates influenza, and sometimes rages under the name of "epidemic pleurisy," or pleuritic influenza." Of the special forms of influenza, "pink eye" is perhaps the best known, for it has been very prevalent of late years. By some it is regarded as a distinct disease, but it is in the writers' opinion most probably only a modified variety of simple influenza. It is usually characterised by the pink colour of the white of the eyes, and by great severity of constitutional symptoms and great prostration. One or both lungs often become diseased, and violent diarrhœa and colicky pain, indicating disease of the organs of the belly, may supervene ; and in some cases there is total loss of power in the hind quarters. This form of influenza is especially dangerous in weakly animals.

The complicated forms of influenza now recognised are three. The first or the "thoracic" form, attacks the organs of the chest mainly, *i.e.*, the lungs, pleura, and heart. The second, or the "abdominal" form, or so-called "bilious fever" or so-called "typhoid fever," attacks the organs of the belly, and is characterised by great prostration, and by the yellow colour of the . membrane of the nose and other visible mucous surfaces, and of the white of the eye. The third and last form is termed the "rheumatic variety." This form usually appears towards the close of the ordinary symptoms of the disease, and is recognised by the pain in the joints, which crackle when moved, and after a time swell. The heart is often attacked in this variety. One point more we should mention as very common in influenza, and this is the great liability for the glands of the throat to become inflamed and swollen, and even to form abscesses.

Influenza is a very debilitating disease in horses, as it is in man, and it is to be borne in mind 'that one attack does not secure immunity from a second or even a third.'

We will now turn our attention to the consideration of the general and special management of this contagious malady. In every outbreak of influenza it should be our first object to inquire carefully into the hygienic arrangements of our stables ; to see that the drains and ventilation are not at fault ; to ascertain—and this is very important—whether or not the water supply is contaminated with sewage matter ; and, finally, to see that the oats and other fodder are in good condition.

If these matters are carefully attended to, influenza will be much less likely to break out in a severe and lingering form than among animals subjected to defective sanitary arrangements.

In speaking of the treatment of influenza, we should always remember that grave symptoms may in many instances be prevented by early and judicious care and treatment. Indeed, it is certain that many a neglected case proves fatal, which with proper care and attention would have proved but a simple, mild attack.

After we have either satisfied ourselves that the hygiene of our stables is not at fault, or have carefully attended to them when defective, we should proceed to isolate the infected animals, and to disinfect the contaminated stalls. As is well known, infectious diseases are so termed owing to the fact that animals affected throw off either in their breath or from their body, or from both, the poison of the malady. The poison, or rather the germs of the disease, consist in all probability in this, as in other infectious diseases, of living vegetable fungi of very microscopic size. They are volatile, and when they gain access to the system, multiply at an enormous rate. In disinfecting, our object is to diffuse a chemical agent, which by destroying them will remove the infection still lingering in the unhealthy stalls.

We do not purpose here to enter into the life history of these low forms of vegetable life, but must point out that their multiplication and dissemination outside the body can be largely prevented by stringent measures. Thus it is well in outbreaks of influenza to wash the harness and fittings, and to purify the drains with a solution of crude carbolic acid, which can be easily procured from any chemist. One can make a solution by mixing equal parts of carbolic acid and soft soap and adding a sufficiency of boiling water. As a lime wash for disinfecting purposes, half a pint of crude carbolic acid may be mixed with each bucketful of lime-wash for the walls. This method of purifying the stable is a very useful one. One of the most powerful volatile disinfectants which we have is chlorine gas, which may be generated in the following manner :— take two pounds of chloride of lime, in an old basin or earthenware pot, remove all the horses, shut all windows and casements, add four ounces or strong oil of vitriol, and stir the mixture with a stick, taking care not to inhale the gas. Then quickly leave the stable and shut the door. In four or five hours the windows and doors may be opened, and after the escape of the gas, the stable may be entered, when it will be well purified from noxious germs of disease. Some authors prefer sulphurous oxide gas, which may be generated by placing sulphur on burning embers in an earthenware pot. The same precautions should be taken as in the case of chlorine gas.

In mild forms of influenza but little medicinal treatment is generally necessary. The animal should be placed in a comfortable and not draughty loose-box. He should have careful nursing, and be seen, except in mild cases, at least once every alternate day by the veterinary surgeon.

We will now say a few words as to the medical treatment of mild cases of influenza. Frequently has the writer heard owners of horses and stock declare that when they have influenza among their horses they continue to work them until well. This is a very great and very serious mistake. Numbers and numbers of neglected cases of influenza have proved very troublesome and fatal in consequence of such carelessness. Only a short time ago a gentleman had a case of influenza in a valuable six-year-old cart stallion. He administered an overdose of aloes, and thought, no doubt, that the animal would soon be well. On the contrary, he became much worse, and the writer was called in. The lungs were found to be diseased, and water in the chest was diagnosed. Two days afterwards the animal died, and four gallons of fluid were found in the cavity of the chest. It is well to bear in mind "a stitch in time saves nine." Rest, warm clothing, the application of bandages to the legs, and hand-rubbing, are all of great benefit in influenza. The diet, should, in the first instance, be of a laxative nature, consisting of linseed or oatmeal gruel, bran mashes, carrots, turnips, or fresh grass. If the bowels are costive, two or three drachms of aloes may be administered. Never give a large dose of opening medicine in influenza, for the mucous lining of the bowel is always more or less irritable in this disease, and is easily excited to undue action, which in all these cases retards cure. For the first two or three days a drench, consisting of liquor ammonii acetatis four ounces, spirit of nitric ether (the best) one ounce, and water to make half a pint in all, may be given three times daily. If the throat be sore, it may be rubbed twice daily with camphorated oil, for which the following is a useful form :—of camphor half an ounce, methylated spirit one ounce, solution of strong ammonia two drachms, olive or linseed oil to half a pint.

In severer cases where the throat is badly affected, a blister of cantharides oil may be applied externally, the head of the animal being tied up ; or, instead of the oil, a hot poultice of linseed and bran may be substituted, and will be found very beneficial. After about three days tonic drenches should be substituted for the fever medicine. The following is a useful form:—of citrate of iron and ammonia two drachms, of aromatic spirit of ammonia one ounce, of tincture of ginger one ounce, of tincture of gentian one ounce, of water or beer to a pint. This may be given twice daily, and will quickly restore tone to the system, and generally give the animal a good appetite.

In conclusion, we must carefully warn our readers against advertised nostrums, whose virtue is to cure all diseases. Confidently we can state that, although in some cases they may do good, and although in other cases one may think they do good, yet nevertheless they bring the veterinarian a very large amount of extra work in the course of each year.

No medicine, as we all know can cure every disease, and we have found that many of the quack preparations, when analysed, contain not only drugs of very inferior quality, but also those which are absolutely deleterious.

Of the disastrous results produced in many valuable studs of horses by the internal use of preparations of aconite, we shall speak when we come to deal with the subject of poisons.

STRANGLES OR COLT-ILL.

MOST of our readers who have had much to do with young horses have doubtless some little knowledge of strangles or colt-ill ; yet, although this malady has been recognised from early times, there are still, even among professional men, many diverse opinions regarding its nature. Strangles is widely spread over the surface of the earth. It is, however, said to be somewhat rarely met with in Southern Europe ; but in Germany and the northern and western countries it is a common malady. In Africa and in Arabia, and indeed in all countries in which the eastern horse, with its nervous temperament is the only representative of the race, it is said to be absent.

Although strangles is more especially liable to attack horses from two to five or six years of age, it may nevertheless infect animals of any age or breed. We must remember that when the disease affects foals it does not confer immunity against a second attack, and when it invades old animals it does not differ much from an ordinary cold with or without soreness of the throat.

Some authors regard strangles as an ordinary catarrh or cold, and others look upon it as allied to glanders and other malignant diseases. We are of the opinion of those who regard it as an eruptive fever, *i.e.*, a fever accompanied by the appearance of an eruption, which in this malady consists of one or more abscesses formed between the branches of the lower jaw. This, moreover, was the view held by the late Mr. D. Gresswell, F.R.C.V.S., who was as satisfied of its contagious nature as he was of that of influenza. Regarding the causes of strangles we cannot speak definitely, yet it is certain that this disease, like influenza, is more especially liable to attack animals which are subjected to unhealthy conditions, such as bad drainage, contamination of the water supply with sewage matter, defective ventilation, overcrowding, and other such like agencies. It is more likely to attack horses debilitated from any cause than stronger and more vigorous animals. Moreover, when it does attack weaker individuals it is more prone to assume a severe type. As an occasional predisposing cause may also be mentioned, the replacement of the milk teeth by the permanent teeth which succeed them. Again, it has been said than any injury may induce a form of inflammation which may precede the manifestation of the malady in question. An insufficient quantity of food, changes in the place of abode, and changes in the weather, are also doubtless of much influence in causing the appearance of strangles in horses, and it has been noted that, like influenza, this disease is more prevalent in spring and autumn than in summer or winter.

Dr. Fleming in his "Veterinary Sanitary Science," says, of horses affected with strangles and sent to the Alfort Veterinary School, that 88 per cent. were found to be newly purchased and imported from breeding districts; and that Reynal has seen six hundred remounts, hurriedly purchased in foreign countries and sent to the army corps, suffer without exception from strangles. The same has happened in remount depôts and regiments receiving new purchases, the disease appearing within a month of the arrival of the animals. Strangles received its name from the great difficulty of breathing, accompanied by a trumpet-like sound, more marked while taking the inspiration, these early signs resembling those of strangulation.

Before speaking of the symptoms of strangles, we ought to point out that there are two varieties of this malady, the one being called mild or benign, the other malignant strangles.

In the benign form the attendant usually first notices slight cough and difficulty of swallowing, owing to soreness of the throat. The animal is dull and dejected, and disinclined to eat. The number of the beats of the pulse is raised, and the temperature is elevated. The breathing also is quickened. In the neighbourhood of the throat there is swelling, and this may occupy the space between the branches of the lower jaw, or it may be formed on one side of the throat only, or it may be more generally diffused among the tissues. The swelling gradually enlarges, and eventually one or more abscesses are formed. When the throat becomes, as it sometimes does, more severely affected, the structures forming the upper part of the wind-pipe become so much swollen that breathing becomes difficult, and thus while the horse breathes there is caused the trumpet sound, due to the passage of the air through the swollen and therefore obstructed air channel. This sound is more marked while inspiring air than while expiring it, and it usually passes off under treatment. Sometimes this inflammatory condition of the throat is more severe and more persistent, and is a source of danger to life by suffocation or strangulation. The benign form seldom extends beyond a period of fourteen days, and nearly always terminates favourably.

In the malignant or irregular form of strangles the fever is more severe, and the glands which secrete the saliva become affected and may suppurate, forming abscesses. Sometimes the glands in distant parts, such as those of the belly form abscesses, and the fever may last a very long time. Malignant strangles may be of one or two months' duration, and the animal may die of various complications, viz., from poisoning of the blood by the formation of matter, or from inflammation of the lining membrane of the belly. Sometimes roaring or whistling is left after recovery from strangles, but the most common sequel of bad attacks of this disease is blood poisoning, resulting in the formation of abscesses in various parts, more especially in the limbs. In speaking of the treatment of strangles, we might repeat with advantage much of what has already been said concerning influenza. In all cases we should endeavour in the first place to isolate the infected animals as quickly as possible, and to keep a careful watch over the others. According to Charlier bovine animals may become infected with strangles. This observer says that he has known of its being so transmitted to beasts when lodged in a badly ventilated filthy stable in which the diseased horses had been kept. Dr. Fleming says that he remembers some years ago a particular stable in the cavalry barracks, at Edinburgh, which was called the strangles stable. The erroneous notion was then prevalent that it was necessary for all horses to have the disease. Remounts were consequently always lodged there, in order that they might become affected. The desired result was nearly always attained. In these days of scientific enlightenment no intelligent person we feel sure can any longer hold such views as these.

We must remember that when an animal takes the infection of strangles, he does not immediately show symptoms of illness. On the contrary, the disease first remains latent or, in other words, in an incubatory condition for a period of one to three weeks or so.

Having isolated the diseased animals, and having placed them under proper hygienic management, we should proceed at once to disinfect the contaminated stables in the manner we have already indicated while treating of influenza. This is very essential, and should never be neglected by those who wish to have their stables healthy. The diseased animal should be placed in a moderately warm, well ventilated, but not draughty, loose-box. The diet should be at first laxative, consisting of such food as mashes, linseed cake, gruel, roots, and grass. In mild cases a febrifuge draught may be given three times a day. The following is a useful formula: take of liquor ammonii acetatis four ounces, of nitric ether one ounce, of bicarbonate of potassium half an ounce, of water to half a pint or a pint.

Should severe throat symptoms supervene, it is necessary to allow the horse to breathe the vapour of hot water. When symptoms of strangulation show themselves, it is sometimes necessary to make an opening into the windpipe, and to insert an instrument through which the horse can breathe air—the inflamed and thus obstructed orifice of the air channel, or glottis, no longer allowing of the passage of air in sufficient amount.

TRACHEOTOMY TUBES.

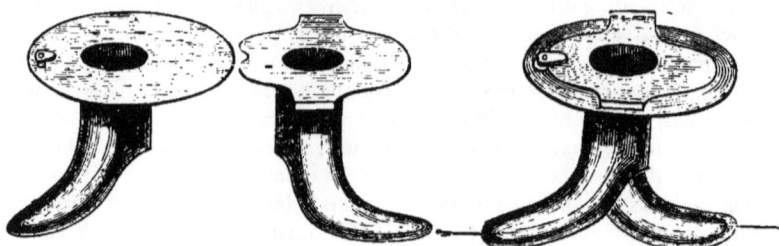

PEUCH'S TRACHEOTOMY TUBES.

The veterinary surgeon inserts the tracheotomy tube in an opening which he makes at about the junction of the upper and middle thirds of the trachea or air passage. This air passage is composed of rings of cartilage or gristle. Two or three of these rings are divided. Then the tube is inserted into the orifice which has been made, and tied in its place. When an abscess is being formed, it is well to apply blisters if it is tardy in coming to a point. Some prefer the application of poultices. Many accidents of a serious nature have occurred from the opening of abscesses by amateurs. In one case under our notice, the duct or tube which conducts the saliva from the gland to the mouth was divided, and for a long time afterwards a pint or more of saliva ran daily from the opening made.

Sometimes little vesicles or blebs form on the skin in parts where a deep abscess is being formed. They sometimes act as valuable guides in helping us to locate points of inflammation. After an abscess is opened it must be kept very clean, and must be dressed with a solution of carbolic acid (1 in 25 of water) or other antiseptic. Such ointments as are called digestive are in certain cases very useful.

In malignant strangles antiseptic medicines, taken internally, are very valuable. We may mention sulphite of sodium especially. These cases are of too severe a nature to be undertaken by other than professional men.

As the fever of strangles abates, tonics and stimulants are required to restore strength; and they are required where the animal is much debilitated, even very early in the progress of the malady. The form given in influenza as a tonic mixture will likewise prove valuable in this disease.

GLANDERS AND FARCY.

GLANDERS was described in very early times by Aristotle and Vegetius, and we read of it as far back as the time of Constantine the Great. It is said to be absent in Australia and rare in India, not breaking out unless it be imported from other countries. It is, for the most part, a disease of temperate climates, and is well known in Norway and Java, and not unfrequently it breaks out at the Cape of Good Hope. Dr. Fleming has witnessed the ravages caused by this dread malady in Northern China as well as in Shanghai.

Glanders is a highly contagious and malignant fever, which, though especially affecting the horse tribe, is also readily transmissible to man, sheep, goats, felines, and rodent animals, as rats and mice. Cattle, pigs, and fowls fortunately cannot be inoculated with the poison of this awful disease.

Glanders, of all diseases to which the horse is liable, is at once the one most peculiar to the equine tribe, and at the same time the one most justly dreaded. It may break out in four different forms : acute glanders, chronic glanders, acute farcy, and chronic farcy, and may assume very different degrees of severity. In the Crimean War glanders broke out in a very fatal and malignant form, and caused very serious ravages among the horses.

Of all the causes which predispose this noble animal to these various forms of glanders none are more potent than defective sanitary conditions, such as overcrowding, insufficient ventilation, bad drainage, and bad general management. It is well known—indeed it has been abundantly witnessed, that horses crowded together in camps or on board ship during long voyages, are especially prone to attack by this disease, owing to the deficient ventilation and want of fresh air. Nothing is more poisonous to any animal or man than breathing over again air vitiated by his own exhalations. How often have we read of the numbers of victims in the days of the slave trade from overcrowding on board ship, and the case of the Black Hole at Calcutta is familiar to every one. Out of the 146 prisoners, 123 died in one night, and several of the survivors afterwards succumbed to putrid fever. It should always be remembered that a due supply of fresh air is quite as necessary in the case of animals as in man for the preservation of a healthy condition.

Debilitating influences, such as old age, bad food, over-work, and lastly exhausting diseases, also predispose the horse to the fatal malady in question.

For many years glanders was supposed to be capable of spontaneous origin in the horse, and many able authorities have written in support of this view.

Regarding its spontaneous origin, we ourselves are not persuaded. It is certain that the disease can spread from one animal to another by actual contact or through the medium of the air. There is no doubt that defective hygiene and violation of the ordinary laws of health render the system a fit receptacle for the development of glanders.

In various parts of our work we shall often have occasion to speak of the so-called "contagium" or "virus" of different diseases, and we therefore propose here to give our readers a very short account of what is meant by these terms.

For many years scientific men have been seeking to discover the nature of the poison or contagium of the specific fevers of man and animals. Each one of these fevers runs a more or less definite course, and presents special characteristics of its own by which it is recognised. The poison of each multiplies in a most marvellous and rapid manner, and one diseased individual may spread the fever among countless numbers in a very short space of time. Even so long ago as the great plague of London, the belief was expressed that the pestilence was probably due to some living organism which entered the blood and rapidly multiplied there, and that it was capable of passing from one individual to another through the medium of the air or by actual contact.

Now, at length, we know that this belief was fully justified, for many of these contagious fevers, both in man and animals, have been proved to be due to the growth and development of poison, which consists in germs of a vegetable nature of very simple structure belonging to the order of the fungi. This theory, which was really, however, first started in 1840 by Henle, was taken up by Schwann and others and perfected by M. Pasteur. It is now accepted by most scientists of the present day, and is termed the "germ theory" of disease. This is one of the most important discoveries of the age in which we live. Although many years ago some eminent authors expressed their belief that there was a living germ, which by its growth caused glanders, it was not until the year 1882 that Schutz and Löffler discovered a low form of vegetable life in glanders. It is called the glanders bacillus, and can only be seen by using a very high magnifying power indeed. It is said to be very like that little but most destructive germ which was found by Koch in persons suffering from consumption of the lungs. Indeed glanders has much resemblance to consumption or tuberculosis of man and animals.

Now it has been found that glanders is developed when the diseased matter is taken in the water or in the food, or even if the diseased material be given in the form of a ball.

It may also be given by inoculating an animal, or by injecting the poison of glanders into the blood of a healthy individual. It should be remembered that the poison when it enters the system does not produce symptoms of disease at once, but remains inactive for a period of from three to seven days or more.

Pus of a Pulmonary Abscess in a Horse dead of Glanders.
1. The nuclei of pus cells.
2. The glanders-bacilli.
Magnifying power 700. (The preparation had been stained with methylene-blue.)
After Klein.

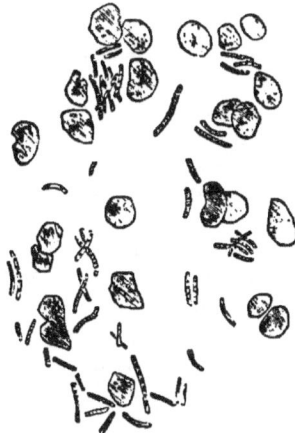

From a Preparation of Human Tuberculous Sputum, stained after
the Ehrlich-Weigert Method.
The nuclei are stained blue, the tubercle-bacilli pink. Magnifying power 700.
After Klein.

We must firstly turn to the consideration of the symptoms of acute glanders. This malady appears suddenly, being ushered in with severe shivering fits. The temperature rises to as high as 106° F., or even higher. In health it should only reach 100°.5. The pulse and the breathing are much quickened, and the membrane lining the inside of the nose is greatly inflamed, varying in colour from a light to a dark brownish coppery hue. In a few days the fever abates somewhat. It again becomes more severe after this short remission, during which the membrane lining the inside of the nose becomes studded with small tubercular nodules arranged in groups, or more generally diffused over the surface. These little nodules vary in size from that of a small seed to that of a pea. In a few days they soften and become converted into ulcers, and then there issues from the nostrils a foul blood-stained fluid. The glands under the jaw enlarge and soon form abscesses and burst. Acute glanders is rapidly fatal, and when the lungs are much affected, death generally occurs before the lapse of three or four hours.

Chronic glanders has special characteristics distingushing it from the acute form. It may continue for many months without obviously affecting the general health of the animal. It mainly differs from the acute variety in that, while the constitutional symptoms are more trivial and variable, the local changes in different parts are more important and numerous. The chronic form is more liable to develop constitutional symptoms at the later than in the earlier, stages. Usually there is a discharge from one or both nostrils, and this is never absent when nodules and ulcers are formed in the lining membrane of the nose. The discharge, at first like that of common cold, gradually becomes thicker and pasty, and has a tendency to adhere round the nostrils. The glands under the jaw enlarge. They are somewhat painful and tender, but gradually become hardened and fixed to the jaw and distinctly nodulated. Chronic glanders is liable to develop acute symptoms suddenly, especially during the later stages.

Farcy is a form or variety of glanders. It occurs as a result of inoculation with the diseased matter of glanders or farcy, or from infection, and according to some it may arise *de novo*, as the result of debilitating influences and bad sanitary conditions, which certainly predispose animals to attack.

In acute farcy there is shivering and a rise of temperature as in glanders, but this form is especially characterised by local swellings, generally confined to the extremities. At first a limb becomes enlarged, hot and painful, and there is marked lameness. When the swelling subsides "farcy buds," or little lumps, and enlarged veins and cords are left behind on the limb. These little buds afterwards become ulcers, and discharge a blood-stained creamy fluid, and the veins and cords may also ulcerate. Sometimes in these cases acute glanders is developed. Now we must carefully distinguish such cases as these from those called "weed," which is of a totally different nature, and in which such buds and ulcers are not formed.

Chronic farcy differs from the acute form only in intensity and duration. It is a very common type of the disease, and is more amenable to treatment than the other varieties. The special characteristics of it are local growths. Nodules are formed in various parts where the skin is thin, as over the face, jaws, throat, and along the neck, forearm, and flank, and they afterwards soften and discharge an unhealthy fluid.

Finally, we must say a few words with regard to the treatment of the forms of glanders and farcy. Animals affected with any of the various forms are in the interests of the community destroyed, according to the provisions of the Contagious Diseases (Animals) Act, in order to prevent the further spread of this loathsome disease. Mild cases of chronic farcy are sometimes isolated and treated medicinally. There is in some cases difficulty in settling the true nature of the disease. The attendant must be very careful not to contract this complaint, as there are a great number of instances where men have been attacked with glanders. As a rule this disease proves fatal, but some cases of recovery both in man and animals are recorded.

In outbreaks of glanders, the infected animals should be at once isolated and professional aid called in. The stables should be thoroughly disinfected with sulphurous anhydride gas or chlorine, as was described in treating of influenza; and the walls should be scraped, washed, and cleansed with limewash, containing one pint of crude carbolic acid to the bucketful. The harness and fittings and other articles which have been in contact with the animal should also be thoroughly cleansed. In these cases an inquiry should be made into the cause of the attack, and the sanitary conditions, if defective, should be forthwith remedied.

Internally iodine and antiseptics have been proved to be the most valuable of all medicinal agents in treating cases of glanders. In conclusion, we ought to mention that glanders is of far rarer occurrence now in this country than formerly, and it is not improbable that as the laws of sanitation become more widely known and respected, this loathsome pest will ere long be a disease of the past. It has sometimes unfortunately happened that glanders has spread rapidly before its true nature has been recognised. Some forty or more years ago the late Mr. D. Gresswell was called to an outbreak of disease among a number of cart horses. They had already been treated by the local veterinarian for nasal gleet, but on Mr. Gresswell's advice as to the true nature of the malady they were all destroyed and buried. Some days afterwards the owner, believing that his horses had been unjustly condemned, threatened to bring an action against Mr. Gresswell for the whole amount of the value of the horses. One of the animals was exhumed, and a very small portion of its blood was injected into the tissues of a donkey in order to settle the question. Seven days afterwards the donkey developed acute glanders in its worst form and died, and thus Mr. Gresswell's action was justified. The stables were then thoroughly disinfected, and the disease stamped out.

ANTHRAX OR CHARBON.

Anthracoid Diseases; Glossanthrax, and Anthracoid Angina.

OF all diseases which affect man and beast, no one is of greater interest or importance than this most destructive scourge. It attacks all animals, including birds and even fishes, and no clime can be said to be exempt from its ravages. Although not often met with in the horse in this country, it is of common occurrence in the equine tribe in Central Hindoostan and in Southern Africa; and is, unfortunately, only too well known to the stock-breeder, in some parts of England especially, and on the Continent, as the so-called "splenic fever" of beasts. It is, moreover, believed to be closely allied to "black leg," "quarter ill," or "black quarter," which malady is especially prevalent among young animals. Anthrax also attacks sheep and pigs, and is not uncommon among other animals. In dogs also it is occasionally met with in those which have partaken of the diseased flesh of creatures which have died of anthrax.

In man, anthrax is often derived from cases of splenic fever of animals, and it is known as "woolsorters' disease" in this country, and as the "Siberian plague" in Northern Asia. In the human species anthrax may also occur as the so-called malignant pustule, which is developed as the result of local inoculation from handling infected wool of animals which have died of anthrax, or from the contact of an injured surface with the diseased carcases.

Although anthrax is now of rare occurrence in the horse in this country, it frequently raged as a malignant and destructive epidemic in man and the domesticated animals in past times. This disease was known at a very early date. It is mentioned in the scriptural records as the "grievous murrain and blains" which affected man and beast in the days of the captivity of the Israelites in the land of Egypt, and we read that the murrain was then upon the horses, asses, camels, oxen, and sheep.* Anthrax was also described by old Greek and Latin writers. The former termed it anthrax, which signifies a burning coal. The seventeenth and eighteenth centuries were especially remarkable for the devastations made by many severe outbreaks of this plague. In 1617 the malady was of so fatal a type that over 60,000 people died around Naples alone from eating of the flesh of animals which had died of the disease. At the present day anthrax often rages as splenic fever in Siberia. As Loodianah disease it is of more frequent occurrence in Central India, and is well known in Australia as the Cumberland disease. In South Africa it is spoken of as the Cape horse sickness, and it is also met with in North and South America. As "Texas fever" in the United States it is of common occurrence, and makes serious ravages among the cattle. According to Toussaint, animals of the value of 20,000,000 francs die annually of splenic fever in France.

Of late years our knowledge of the nature, causes, and methods of prevention of anthrax has, owing to the labours of scientists, been considerably increased; yet it is remarkable that—although in France and Germany matters are different—in England there is, as far as we know, no enactment which enables anyone to interfere with the disposal of carcases of animals which have died of the disease, nor are there any specified regulations regarding the drainage of lands on which splenic fever appears periodically.

Regarding the causes of anthrax, we may mention that it is especially prevalent in low-lying, swampy districts, where the soil is rich in organic matter, these conditions being in the highest degree favourable for the

---

*Exodus, c. ix., v. 3.—"Behold, the hand of the Lord is upon thy cattle which is in the field, upon the horses, upon the asses, upon the camels, upon the oxen, and upon the sheep: there shall be a very grievous murrain."
Exodus. c. ix., v. 10.—"And they took ashes of the furnace, and stood before Pharaoh; and Moses sprinkled it up toward heaven; and it became a boil breaking forth with blains upon man, and upon beast."
Vide also, Deuteronomy, c. xxviii., v. 27 and 35.

development of this disease. Anthrax is frequent in morasses and in countries exposed to inundations, and in places where water stagnates on the surface of the soil

VIEW OF A SITE ON THE WOLDS OF NORTH LINCOLNSHIRE
in which anthrax broke out periodically; now drained thoroughly.

Indifferent diet, such as fermenting grains, or damaged food ; defective sanitation, as bad drainage and ventilation ; or food and water contaminated with the germs of the disease, are also potent causes of anthrax.

In the spring of 1884 an outbreak of anthrax occurred in a number of cart-horses under the care of Messrs. Leather, of Liverpool. These animals had been fed for some time previously on an Indian pea, which in reality, however, is more like a lentil than a pea. It is imported into Glasgow from India, and has been given not only to horses, but is also used when ground or mixed with some kinds of cake for cattle. The horses attacked commenced to die very suddenly some time after the Indian peas had been given, and for some weeks after their use was discontinued they still died. The symptoms were roaring, bleeding from the nose, great prostration, swelling of the throat followed in many cases by sudden death. When the Indian peas or "mutters" as they are termed, were examined they were found to be very dirty and dusty, and among the dust the germs of the disease were found. Animals have died on the Continent with similar symptoms, after being fed on the legumen, "Lathyrus sativus." It therefore remains to be shown whether their death was caused by a poison in the leguminous plant, or by the bacilli entering the system.

c

Delafond in the year 1845 first showed the vegetable rod-like bodies or bacilli anthracis peculiar to anthrax. In our last article on glanders we spoke of the so-called germs, and showed of what great importance they were in many diseases. They gain entrance into the body by direct inoculation, or through the mouth, or through the air passages. These bacilli of anthrax (of which we append drawings) are now almost universally believed to be the actual exciting cause of anthrax. Either these bacilli themselves or their spores are always present in enormous numbers in the blood of animals suffering from anthrax fever or splenic fever. The bacilli may be separated and washed with distilled water, alcohol, and ether, and dried, yet they still cause anthrax fever when introduced into animals. Pure cultivation of this germ through fifty generations may be made with the same result. The germ always gives rise to anthrax fever, and never to any other. Therefore in it we recognise the direct cause of the malady.

HEART'S BLOOD OF A MOUSE DEAD OF ANTHRAX.
1. Blood-discs.
2. White blood-corpuscle.
3. Bacilli anthracis.
Magnifying power 700. (Fresh specimen.) After Klein.

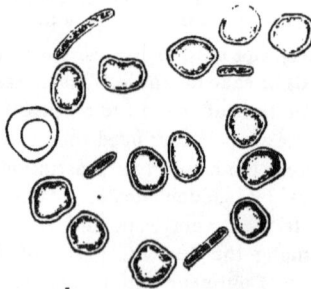

BLOOD OF A GUINEA-PIG DEAD OF SYMPTOMATIC ANTHRAX.
Blood-corpuscles and between them several bacilli.
Magnifying power 700. (Stained with Spiller's purple.) After Klein.

Many outbreaks of anthrax have been traced to the indiscriminate burial of carcases of animals which have died of the disease. According to Pasteur, spores of the bacilli are brought to the surface of the earth by earth worms, even ten to twelve months after burial of the diseased carcases; but this method of propagation is doubtful. Dogs, after feeding on the diseased flesh, may bite sheep, and thus inoculate them with the disease. Flies feeding on anthrax blood have been shown to absorb sufficient poison in their proboscides to give the disease to animals. Anthrax may also be spread by eating food contaminated by water containing the germs of the disease. Uncleaned knives used in dressing carcases may also propagate anthrax. The germs are said to be most volatile in spring and summer, less so in autumn, and still less in winter.

Although anthrax has not yet been shown to be infectious, yet there is every probability that it is. It often appears first in districts where it is local or endemic, and afterwards assumes an epidemic form, spreading rapidly and infecting many animals. We must now turn to the consideration of the symptoms of the forms of anthrax in the horse.

Anthrax fever in the horse is rare in Great Britain. The symptoms appear suddenly. The animal trembles violently, perspires very freely, and breathes irregularly and with difficulty, loses control over his movements, staggers, and dies convulsed. These acute symptoms may however abate in two or three days, and finally end in death or recovery. Sometimes they are not so rapidly manifested. A disinclination to move, a loss of power over the muscles, drowsiness, stupidity, and great prostration, are marked features. Other symptoms are a yellowish bloody discharge from the nose, fetid breath, and pain in the body, which last is shown by uneasiness, pawing, and frequent looking at the sides. The excrement is fluid and blood stained, the skin is harsh and dry, and in some cases crackles when pressed towards the loins. The pulse is irregular, increased in number, and the temperature is raised several degrees, and may reach 106° to 108° F. The respirations become tumultuous and hurried, and the nostrils widely opened; the temperature falls, and the animal staggers at every step; convulsions and delirium ensue; and death ends the scene. Sometimes in the horse, at a certain stage in the fever, there is an external eruption of tumours called "anthrax pustules." They are especially met with in the upper part of the throat, the lower part of the neck, on the back, and in the groin. As the tumours develop, the fever often abates, but in rare instances when they vanish the disease assumes all the characteristics of anthrax without tumours, the general symptoms reappear, and the animal dies in about twelve hours.

The horse is also liable to two forms of anthrax, characterised by changes in special parts. These are termed anthracoid diseases. They have not as yet been proved to be due to the bacilli anthracis, and are named glossanthrax and anthracoid angina respectively, and are generally associated together, rarely occurring separately. The tongue in the former malady is swollen and hard, hangs from the mouth, is of a dark bluish or black hue, and is often lacerated by the teeth. The animal has great difficulty in swallowing. Blebs or vesicles form on the sides and surface of the tongue, and the animal often dies in from twelve to twenty-four hours after their appearance. When they burst, after increasing in size, they discharge an acrid fluid, and leave an unhealthy ulcerated surface.

In the second form, which seldom occurs alone, the tissues round the throat become much swollen, and involve the structures of the head, the breathing becomes difficult, and death results from suffocation.

We will now review the general methods of prevention, and treatment of the forms of anthrax in the horse, but may first show our readers the appearance of an anthrax pustule in a man who had been inoculated through handling infected bales of wool imported from China. This case was treated by excision, or removal of the tumour, and the internal administration of sodium sulphite, recommended by the writers. It terminated in complete recovery. (*British Medical Journal*, June 14, 1884.)

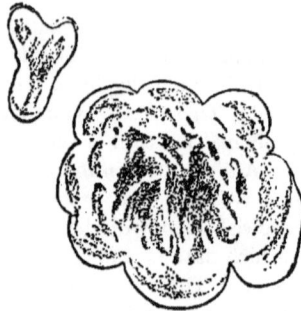

ANTHRAX PUSTULES IN MAN.

One of the main causes of the diminution of the number of outbreaks of anthrax among catttle and horses in this country is the thorough drainage of many of the formerly infected areas.

In all outbreaks we should carefully inquire into the food and water supply, and take special care to see whether there is any escape of effete matter into the wells or ponds. Ascertain if the food be mouldy or fermenting, as mouldy grains have been shown to be the exciting cause in several instances. In the year 1878, the late Mr. D. Gresswell had under his care a large number of cattle and horses affected with anthrax, supposed to have been caused by eating mouldy grains.

The healthy animals should be isolated from those affected. The stables in which the horses contracted the disease or died, and all the implements, such as the fittings, should be disinfected and cleansed. The carcases should be deeply interred, and the litter and manure burned.

Although outbreaks of anthrax fever are rare in horses in this country, they are not so rare among cattle, from which horses sometimes contract the disease. Should the outbreak occur in a particular infected area or field, this should at once be thoroughly drained, and the animals removed from it. The site of which we append a drawing has almost every year been a source of great loss to the farmer holding the land, but it is now thoroughly drained according to our directions, and there have been no further outbreaks since.

In France, cattle are largely inoculated with the so-called anthrax vaccine, which develops the disease in a mild form, and renders the animals in most instances proof against further attacks. Large numbers of sheep and other animals have been thus inoculated with vaccine by M. Pasteur, and the results of his labours attest the practical value of his conclusions.* It is our practice to administer sulphite of sodium to animals which have been in contact with the infected ones, in order to act as a preventive, for which purpose, we believe, it is very effectual ; and it is also of undoubted value in the early stages of anthrax fever, and possibly of some use even in the more advanced conditions of this most fatal disease.

Some authors prefer the internal use of carbolic acid and other antiseptics ; but we believe them to be less effectual than the sulphite given in two to six drachm doses. The late Mr. D. Gresswell, examiner in cattle pathology at the Royal College of Veterinary Surgeons, was also of this opinion, and Mr. Charles Gresswell, of Nottingham, also concurs with this view.

Of the treatment of glossanthrax and anthracoid angina we need say but little, as these diseases are so often fatal, no matter what treatment be adopted. The internal treatment is the same as that for anthrax fever. The vesicles on the tongue may be opened and dressed with carbolic acid solution (1 in 30 of water). Fomentations of the swellings are beneficial. Where the swelling is causing suffocation, tracheotomy is necessary. Professional advice is called for in all outbeaks of anthrax, not only for the sake of the general management, but also in order that proper steps may be taken to stamp out the disease.

SCARLET FEVER AND PURPURA.

THE term blood poisoning, although an ambiguous one, owing to its being applied to several different diseases of the horse, is nevertheless a convenient appellation for those two fevers :—scarlet fever and purpura, which are accompanied by the formation of an eruption.

The more scientific term for these two closely allied diseases, with the account of which ends our description of the so-called zymotic fevers of the horse, would be "the eruptive fevers," or fevers in which a definite eruption breaks out in the skin and in the membranes lining the nose and mouth. The cause of these two diseases, which, although presenting great

---

* According to Dr. Klein, animals inoculated in this manner are only protected against the disease for a term of nine months or so.

resemblance, are nevertheless quite distinct, are very similar. In almost all instances they break out after or during some weakening disease, more especially influenza and strangles, and are in almost all cases traceable to bad drainage or insufficient ventilation, or to both these causes. Our readers will no doubt easily understand that when animals suffering from influenza or other debilitating diseases, such as strangles, are · closely confined in ill-ventilated and badly-drained stables, and are made to inhale the products of their own excreta, and to breathe over and over again the air contaminated by the exhalations of their bodies, they become still more weakened, and fall a ready prey to these eruptive fevers. Purpura and scarlet fever very rarely occur as primary diseases, but, as we have said, nearly always follow some other debilitating disease, and in most instances their causation depends upon bad hygienic conditions.

Horses sent to work too quickly after attacks of influenza and strangles not unfrequently develop purpura or scarlet fever in consequence of the strain put upon them in their enfeebled condition. Again, in some instances, purpura or scarlet fever breaks out in previously healthy horses. merely as the result of bad hygienic conditions, and the non-observance of the ordinary rules of health. We have had many severe cases of purpura in cart horses ; but in most instances the disease followed influenza, which had been greatly neglected and carelessly managed.

Sometimes, we must remember, influenza is of a very severe type, and so weakens the animal and poisons the blood, that, even where the hygienic conditions are pretty good, nevertheless scarlet fever develops. In some cases it manifests itself after an attack of influenza in horses whose constitutions are bad.

We will now speak of the symptoms, first of purpura, then of scarlet fever. Usually, in purpura, the first noticeable symptom is the sudden appearance of local swellings in different parts of the body—in the limbs, belly, head, but more especially around the nostrils, mouth, and lower parts of the face. In a severe case under our treatment, the disease began with huge swellings of all four limbs, which were so hot and painful that the animal, a valuable six-year-old cart horse, could not stand for more than a few minutes at a time. Large bluish-black spots of the size of half-a-crown appeared about the end of the nose, and the membrane lining the inside of the nostrils was of a bluish-black hue. Sometimes we may see little purplish patches in this situation, but they gradually coalesce together, and become more darkly coloured. There was a great flow of saliva from the mouth, and a blood-stained serous fluid oozed from the nostrils. These swellings in purpura terminate abruptly, that is to say, they do not shade insensibly away. They are tense, hot, and painful, and are due to the exuding of blood and serous fluid into the tissues. Little blebs of about the size of peas appear in most cases on the lower parts of the limbs, around the hocks and fetlock joints, and after a time they burst and discharge an amber-coloured serous fluid. The pulse in the case mentioned was very feeble, and varied in number from 100—110 beats per minute, and the temperature remained for three weeks as high as 104 F. On the fourth day the swellings began to abate, the pulse fell to 96 ; but on the fifth day the symptoms became aggravated, and the head swelled so enormously, that death seemed imminent from suffocation. Little blebs or " vesicles " formed all over the swellings and in other parts of the body. They shortly afterwards burst, and from them ran a serous fluid. The swelling of the head afterwards gradually diminished in size, and on the tenth and eleventh days the swellings—as they often do—began to disappear from one part, and to reappear in others. In this case the belly and the sheath swelled enormously, and attained a huge size. In the third week immense sloughs of the skin formed in parts which had been swollen, and large unhealthy sores discharging fetid matter were left. The sheath formed an immense tumour, and many pints of serum escaped continually from the sores. Huge pieces of skin became detached, and the flesh adjacent rotted considerably. A piece of skin half a square foot in area came off from the belly, and another large piece rotted away from adjacent parts. Large fragments also sloughed away from the inside of the thighs, and this detachment of patches of skin continued for the space of two weeks. Then, with very careful treatment, the sores gradually healed, and the horse made a good recovery. In most cases of purpura the bowels are confined. Sometimes the excrement is blood-stained, and pain in the belly is frequently manifested. The appetite is impaired or lost during the disease ; and a hoarse, hollow cough is often present. The patches formed on the nose often slough, leaving raw, ulcerated surfaces, and it is from these that the blood-stained secretion runs. The tongue frequently has blebs on its surface ; and the animal moves with difficulty, owing to the stiffened and painful condition of the joints. With great care most cases of purpura recover, but sometimes even a mild case, when first seen, may eventually prove fatal.

We will now turn to the consideration of the symptoms of scarlet fever. This fever usually begins towards the end of the first week of the primary malady. It may, however, begin as early as the third day, or even as la as after the end of the first week. Sometimes, more especially when i occurs as a primary malady, scarlet fever proves so mild as scarcely to affect the general health. An outbreak of such a kind occurred a short time ago in our practice. This form is called simple scarlet fever. In it the pulse is raised to 45—50 beats per minute ; the temperature rise one or two or more degrees; and then, after this rise, a rash appears. Little smooth blotches are developed on the skin of the face, neck, body, and extremities, but the skin itself is not much elevated. Little blebs also form, especially on the inner sides of the thighs. The membrane lining the nose and mouth is covered with scarlet spots or streaks of variable size. The spots are especially seen on that part of the membrane of the nose which separates the two nostrils, and also on the inner surface of the lips. From the nose runs a serous discharge, which gradually becomes yellowish brown in colour. The limbs are generally swollen, and the animal is stiff. In some cases there is no eruption, but little spots appear on the nose. The throat is generally sore. These cases recover pretty rapidly, as a rule, in about twelve or fourteen days. In the severe form of scarlet fever, termed "scarletina anginosa," the throat and upper parts of the air passages are more especially involved. The symptoms are first those of simple form, and then they gradually increase in severity, or they may be severe even at the outset. The limbs swell rapidly. The rash and the blebs are more often found on the limbs than on the body, and they appear in successive crops, and often spread by coalescence of neighbouring patches. The spots in the nose and mouth are larger and darker in colour than in the mild form. There is great·difficulty in breathing and swallowing, and loud and painful cough in many cases. The throat is much swollen, and sometimes, though rarely, abscesses form in the swollen glands below the jaws. The pulse rises to 60—100 beats per minute, and is very weak. The breathing is much quickened, and the bowels are confined. This fever, when not complicated, generally declines in six to eight days ; but great debility often remains after the acute symptoms have disappeared, and there is great weakness of the heart. In these severe cases a fatal result is often to be greatly feared.

The reader will now see that purpura and scarlet fever are very similar. One may distinguish them by the fact that in the latter the spots in the nose are scarlet, whereas in the former they are dark purple. Sore throat is never absent in the scarlet fever, but is rarely present in purpura. In scarlet fever also there is swelling of the glands, whereas in purpura this does not occur. Again, the tendency to sloughing in various parts of the body is characteristic of purpura.

We have now, lastly, to speak of the treatment of these two diseases. In the first place, chiefly, the sanitary conditions should be carefully attended to, and the infected animals isolated in well-ventilated, warm, loose boxes ; for, although there is some doubt as to whether these fevers are contagious or not, there is, nevertheless, some evidence in favour of their being so. Some authorities, we may mention, do not believe them to be contagious. Mr. Charles Gresswell, of Nottingham, has held with us that scarlet fever is, at any rate in some instances, contagious. The stables in which the diseased animals have been confined should be disinfected thoroughly, and the walls well washed with lime-wash, containing half a pint to a pint of crude carbolic acid in each bucketful. The diet should be laxative and nutritious, consisting of linseed cake or oatmeal gruel, and roots. Mild cases of scarlet fever may be treated by the administration, three times daily, of the fever draught prescribed in influenza ; but the more severe forms require all the care of the scientific veterinary surgeon.

In such cases antiseptics are of great value. Fomentations to the throat and inhalations of steam from boiling water are very beneficial in all cases. During convalescence the tonic mixture prescribed in influenza may be given three times daily, and very careful attention should be paid to dieting. The food should be of the most nutritious kind, but not in great bulk at first until the digestive powers are fairly restored. Moderate exercise should be enjoined as the animal regains strength.

In purpura, also, the diet should at first be moist and nutritious. The medicinal treatment of this dangerous malady requires great professional skill in order to bring it to a successful termination. The swellings should be fomented, but not punctured. The sores should be kept very clean and dressed with some antiseptic solution, as carbolic acid (one part in forty of water). During recovery tonics are required.

GENERAL DISEASES DUE TO ERRORS IN DIETING AND MANAGEMENT.

Weed or Lymphangitis, Diabetes Insipidus, Diabetes Mellitus, Oxaluria, Azoturia.

WEED OR LYMPHANGITIS.

UNDER this heading we propose to treat of those general diseases of the horse which are in most cases due to dietetic errors, or to some irregularities in the management. Of the diseases of special parts, such as colic, due to similar causes, we shall speak when we have to deal with local diseases.

The first of the general diseases which we here wish to bring before the notice of our readers is one commonly known in Lincolnshire as "weed," "the humour," "farcied leg" (though it has no relationship with farcy), sometimes spoken of as the "Monday morning disease," and in scientific language termed "lymphangitis."

Of this malady it is very essential that all who keep draught horses especially should have some clear knowledge ; for, with due precautions and careful management, it is, in common with some other general diseases of the horse of which we now speak, largely preventible, and very amenable to judicious treatment. Weed is not at all an uncommon disease of the horse ; but we should mention, before entering into details, that it is a special inflammatory malady, and must not be confounded with other forms of disorder such as humour or farcy.

It is a general affection of the constitution attended by inflammation, beginning in the glands at the upper part of one of the limbs, in most cases a hind one. The leg becomes swollen, and when pressed upon by the finger "pits." In some cases both hind limbs are affected, and in rare instances a fore limb is the seat of the disease. The commonest situation however is the left hind leg. Regarding the nature and causes of weed, we may say that it is a general disorder of function, especially associated with impaired digestion and disordered assimilation of the food. It is especially a disease of the heavy draught horse of sluggish lymphatic temperament, and is particularly common among certain breeds of agricultural horses.

One attack renders an animal more subject to a second, and in many cases one seizure succeeds another periodically, until the limb becomes permanently enlarged; and this is the condition which has often been confounded with farcy. The writers have seen cases where a second and a third attack have been followed by a fourth, fifth, and even a sixth.

The chief cause of weed is feeding beyond the requirements of the healthy nutrition of the animal. Cessation or sudden diminution of work in well-fed horses is also a common cause, as is seen in the frequent occurrence of this malady among heavy draught horses after a Sunday's rest. The complaint has in consequence been named the "Monday morning disease." Sudden or prolonged exposure of the horse to cold or damp will in many cases bring on an attack by suddenly disturbing the digestive functions, and indeed any rapid change in the work or habits of the animal may bring on an attack. Lastly, "weed" may occur as a local inflammation of the limb resulting from an injury. For example, an injury to the foot in shoeing may bring on an attack. The inflammation spreads upwards to the groin, and thus differs from the general disease, in which it begins above and spreads downwards. The symptoms of weed, like those of many other diseases vary very much with the intensity of the attack, and though they are quite characteristic, the writer has nevertheless, strange as it may seem, been called on several occasions to cases of "weed" which were being treated by farriers as "inflammation of the lungs." In many cases a shivering fit precedes the local inflammation of the limb or limbs, and this may last during some hours. As a rule the intensity and duration of this "rigor," as it is termed, is a mark of the severity of the attack. At an early stage there is restlessness and lameness, and after the shivering fit a hot stage follows. The fever runs high. The pulse varies from 70—100 per minute, and is hard, full, and firm. Sometimes, though not always, the breathing is also much accelerated, and sweats bedew the body. The quickness of the breathing is the factor which leads so many farriers to treat weed as if it were inflammation of the lungs, a disease which does however sometimes complicate weed. In weed the bowels are constipated, the urine is dark coloured, and the temperature is raised from $2 \cdot 5$ to $3 \cdot 5$ degrees above the normal, which in the horse is $100 \cdot 5°$ F. The appetite is lost, and there is great thirst and restlessness. The swelling is very tender, and rather firm. It is first noticed in the groin, or in the corresponding region of the fore part of the body. It feels hot, and gradually extends downwards, first on the inner side of the thigh, but gradually encircles the whole of the limb. This pain and lameness increase until the crisis of the fever is reached, and then they remain stationary for a day or two. In severe cases a serous exudation often occurs over the inner surface of the limb, and particularly at the "bend" or "flexures" of the joints.

The general and the local symptoms continue to increase for 24 to 48 hours, and then remaining stationary, are followed in a day or two by subsidence of the fever and gradual diminution of the local swelling.

When weed recurs in a limb, there is less chance of complete recovery, for the tissues of the part become augmented, the entire bulk of the limb becomes increased, and the skin is thickened and hardened. This condition, called in popular parlance "farcied leg" (though not allied in any way to "farcy,") is termed "elephantiasis." Sometimes abscesses form after the subsidence of the fever, after a first or a second attack; and they are generally confined to the tissues just beneath the skin.

Not long ago we were called to a very acute case of weed in an eight-year-old cart mare. The near hind leg and the off fore leg were immensely swollen. The pain was most acute, and the breathing was short and quickened. The subject of the attack had received about 1½ pecks of oats with cut straw and 2lbs of linseed cake every day.

The great majority of cases of weed recover completely, but a thickened limb may be left as a testimony to the former attack. Sometimes, as we have mentioned, inflammation of the lungs, and also in rare cases bowel complaints, may supervene in weed.

We will now turn to the treatment of the disease. The animal should be placed in a cool, well-ventilated, but not draughty loose box. In all attacks, if the horse is in good condition and has received plenty of good food and is not aged, we practice bleeding in the early stages; and, indeed some acute cases would probably prove fatal in spite of all internal medicines unless this method of treatment were adopted. In the above case the writer abstracted five quarts of blood from the jugular vein, and in a few minutes the relief afforded was very marked. Bleeding can not be undertaken by the unskilled. Some writers, we should mention, do not recommend bleeding. The late Mr. D. Gresswell invariably practised it in acute attacks in well-nourished animals, and many other authorities are also of the same opinion. Of aloes we administer four or five drachms in the form of a ball in the first instance, and we do not repeat this dose. Every four hours during the fever, a draught containing :—of liquor ammonii acetatis four ounces, of bicarbonate of potassium half an ounce, of nitric ether one ounce to one ounce and a half, of Fleming's tincture of aconite (in bad cases) five to seven drops, and of water to half a pint or a pint, may be administered.

The affected limb should be fomented with warm water, and, if the pain be severe, tincture of opium may be added to the water, in quantity about one ounce of the tincture to a pint or a pint and a half of warm water, or this may be applied as a lotion after each fomentation.

Fomentations of tepid water should be continued for two hours at a time, four or five times during the course of the twenty-four hours. The diet must be carefully attended to. In the early stages a restricted and cooling diet should be ordered; but in the later stages, when debility supervenes, the food should be nutritous and well regulated. The limb may be supported by bandages applied pretty firmly. When the limb remains much thickened ·after the fever is over, a draught, consisting of iodide of potassium a drachm and a half, of iodide of iron a drachm and a half, and of nitric ether one ounce, may be given in half a pint of water three times daily.

"DIABETES INSIPIDUS," AND "DIABETES MELLITUS."

The next general disease due to dietetic errors is diabetes. Of this malady there are two forms. The first termed diabetes insipidus, is fairly common. The second, termed diabetes mellitus, is rarely met with. Diabetes of the first kind is a malady caused by feeding on mouldy hay, musty, damp, or kiln-dried oats, and bad corn. Sometimes it is caused by boiled food. It may result from prolonged exertion and exposure to cold when the diet is not at fault. Sometimes it comes on from very slight errors in dieting during convalescence from weakening diseases. " Diabetes Insipidus " is known by the great thirst it occasions, and the excessive passing of water and depraved appetite. The mouth has a nasty, sour smell, and the animal gradually loses strength. The treatment consists in change and careful regulation of the diet. At first the food should be restricted to a moderate amount of good hay and mashes, and the animal should be rested and carefully attended to. Mild aperients, such as half a pint of linseed oil may be given when necessary; and half an ounce of bicarbonate of potassium, with a drachm of iodide of potassium may be administered three times daily in the drinking water. Afterwards, during recovery, vegetable tonics, such as ginger and gentian, may be given in the form of a ball, with a drachm and a half of carbonate of ammonium. The other or second form of diabetes is characterised by excessive passing of water, containing sugar in abundance. We need not here enter further into a consideration of this rare and intricate malady.

OXALURIA.

" Oxaluria " is a disease characterised by great debility, loss of flesh, stiffness in the loins, and a branny scurf on the body It is caused by irregular feeding, irregular exercise and work, and indigestion. It is commonly met with in hunters, which work irregularly, and commonly undergo long periods of fasting. " Oxaluria " is also induced by food rich in sugar, such as carrots, turnips, and other roots, especially if the digestion is out of order. In this complaint a purgative should be given and the diet carefully regulated, and roots and other food containing much sugar should be discontinued. A draught, consisting of diluted nitro-hydrochloric acid one drachm, of tincture of nux vomica one drachm, and of tincture of gentian one ounce, may be given in a sufficiency of water three times daily. If the water contains lime in excess, it should be changed. Regular exercise and fresh air are essential.

AZOTURIA.

Lastly, we must say a few words of still another general dietetic disease called azoturia. It is a malady characterised by spasms of the large muscles of the posterior part of the body and limb, and the passage of very darkly-coloured water. This disease is especially apt to follow periods of idleness, preceded by periods of active work, and its primary cause is dietetic. "Azoturia" does not so much occur during actual rest as when the animal resumes work. Tares, vetches, and leguminous vegetables are especially apt to induce this serious disease. It is more common in autumn, and is more frequently met with in mares than in geldings. In no case can the amateur undertake the treatment of this serious disease. The animal should at once be placed in a well-ventilated loose box, with plenty of straw in it, and the diet should be restricted in amount, and of a light digestible kind for the first few days. Recovery is generally the reward of judicious treatment, good management, and careful attention.

LOCK-JAW OR TETANUS.

THERE is no disease to which the horse is subject which is so much and so justly dreaded as is lock-jaw or tetanus, a malady to which the horse and sheep, of our domesticated animals, are the most liable.

Lock-jaw is a grave malady, characterised by continued spasms, not only of the muscles which are under the control of the will, such as, for example, those of the limbs, but to some extent also of the other muscles. These spasms are painful, and from time to time they become more severe and are then followed by intervals of repose.

In most instances, lock-jaw arises in connection with some wound or injury, though sometimes it occurs without any apparent cause whatever. When traceable to an injury it is spoken of as traumatic. When it arises without apparent cause it is termed idiopathic. We must remember that the liability to traumatic tetanus in no way depends upon the severity of the injury, as it not unfrequently follows very slight wounds. It is most likely to follow either punctures or lacerated wounds. Although it has been said by some that lock-jaw is rarely due to wounds of the feet, this is, nevertheless, most certainly an unwarrantable assertion ; for very many cases of tetanus under our care have been due to injury of this most wonderfully constructed mechanism. Wounds of the thighs, feet, quarters, and forearm are especially liable to be followed by lock-jaw, and this is more particularly the case when the nerves are injured. In a case in which a piece of straw was embedded in one of the main nerves of the limb, the late Mr. D. Gresswell found this structure to be in a highly congested condition for some distance from the point of injury. Wounds, in parts which are the most tense, and in structures bound together by unyielding tissues, are more frequently followed by lock-jaw than injuries in the laxer tissues. Injuries of the joints although frequently inducing a high state of fever, are nevertheless not often followed by tetanus. The operations after which this disease most commonly supervenes are docking and castration. In some instances the insertion of setons has been followed by an attack. When tetanus succeeds docking, this operation has in almost all instances been unskilfully or unadvisedly performed under unfavourable conditions, as, for instance, when the animal was in a weakened and debilitated condition, or when after the operation the horse has been confined in damp or draughty stables, and has probably been ill cared for in other respects also. The authors have, moreover, noticed that when docking is performed by means of a blunt instrument in an unskilful manner, tetanus is very liable to follow. When docking has been judiciously performed, we have never known it followed by tetanus.

In some cases tetanus is due to irritation of the stomach and intestines, caused by worms or collections of sand which have been ingested ; and sometimes irritation of the womb following abortion is a cause of this dread malady.

Tetanus, when not due to a wound or injury, is generally traceable to cold and damp, especially after exhaustion. Horses when clipped are sometimes afterwards attacked ; and similarly sheep, when exposed to cold and wet immediately after being shorn, not uncommonly manifest the symptoms of tetanus.

It has been observed that lock-jaw is more prevalent in certain districts than in others ; for in some parts of the country it is very frequently met with, while in others it is as rarely seen. In Lincolnshire it is fairly common. Sometimes it occurs as a local disease, and Professor Williams records that in the summer of 1858, he witnessed ten cases in a fortnight, and of these some were due to injuries, while the others where not traceable to wounds of any kind. In the human species it has been noticed that tetanus is very common in the tropics, apart from any injury. As to the real nature of tetanus there is some doubt. Some hold the traumatic form to be due to irritation of the nerves implicated in the wound, and think that the spasms result in consequence thereof. In support of this it has been argued that many cases of lock-jaw following docking have recovered after repeating the operation higher up. Yet it must nevertheless be remembered that recovery does not invariably follow the repetition of the operation, and moreover, some very mild cases of traumatic tetanus recover without any treatment whatever.

We are of the same opinion with those who view tetanus as a blood disease, and several reasons can be advanced in support of this theory. Firstly, we have seen that tetanus often occurs without any injury. Secondly, tetanus is more prevalent in certain districts than in others, and is sometimes localised to certain parts of the country, affecting several animals at once. There is, however, no proof that tetanus is ever contagious. Thirdly, the resemblance of this disease to rabies or hydrophobia, which has been shown by that eminent pathologist, M. Pasteur, to be due to certain low forms of vegetable life, suggests a similarity in the nature of the cause. Lastly, tetanus is said to be transmissible to man if the flesh of animals which have died of this disease be partaken of. This statement requires confirmation.

Tetanus may be acute or chronic, and there are also several varieties which have received various names from the particular muscles mainly involved.

When lock-jaw owes its origin to a wound, it usually manifests itself in from 10 to 28 days after the infliction of the injury, or it may occur even at an earlier date than this. Generally a stiffness about the neck and lower jaw and of the muscles near the seat of injury is first noticeable. There is difficulty in mastication and swallowing, together with increase of the saliva and a peculiar champing of the teeth. If the head be suddenly elevated, or the horse suddenly turned, there is a characteristic profusion of the "haw" or "membrana nictitans," over the eyes, which are withdrawn into the orbits, thus causing the animal to show the white part of the eye at each convulsive retraction. As the disease advances, the stiffness becomes more marked, especially in the muscles of mastication and in those of the upper part of the neck. The affection soon spreads to the muscles of the body, back, and hind quarters. At length the tetanic condition becomes established, and is very apparent, even to a superficial observer. The limbs are extended and kept apart, the jaws are immovably fixed, the tail is elevated, and the animal moves in a peculiar stiff straddling way, with great difficulty and pain. The pulse is generally not much affected in the first instance, but in a day or two it becomes quickened. In severe cases it may be very rapid in the early stages. Sometimes, also, the temperature rises very high, the bowels are constipated, and during the course of the disease there are periods of calm alternating with violent paroxysms. These latter are easily induced by any sudden disturbance of the animal, such as by loud noises, or by sudden flashes of light into the darkened box. Quietude tends to subdue the patient in a corresponding degree. During the continuance of the spasms, the breathing becomes quickened and difficult; the surface of the body is bedewed with perspiration; the nostrils are dilated, and the nose protruded.

The duration of cases of tetanus varies markedly. Some of the more severe cases run their entire course in less than 48 hours. In other instances the animal may live two or three weeks, and then succumb at the end of that time. As a rule tetanus runs a more rapid course in thoroughbreds than in animals of coarser breed, and appears to be of a more active type in excitable horses than in animals less sensitive. When the animal progresses favourably, the tetanic condition gradually and slowly declines, lasting from three to five weeks.

There is a notion, unfortunately, in some parts of the country, that horses afflicted with lock-jaw invariably die; and the owners, in some instances, refuse to have the animals treated in consequence. This is, however, a grievous error, for, excepting in those instances where tetanus is so acute at the outset that a fatal result is certain, a favourable termination is by no means uncommon under judicious care and treatment. Indeed, in all cases where the symptoms come on slowly, and the animal is able to take nutriment, and lives to the eighth or ninth day, recovery is fairly common.

The animal should be placed in a large, well-ventilated, well-bedded, loose box, which should be kept dark. *Strict quietude should in all cases be enjoined. It is of the utmost importance in the treatment of this affection.* The wound, if there be any, should be carefully examined, and all irritating matter washed away. In cases following docking it is considered by eminent authorities to be advisable to repeat the operation, and where this has been done, in a large number of instances statistics show a large percentage of recoveries. In almost all instances slinging is necessary in the early stages of the disease.

Among the many drugs which have been advocated for the cure of tetanus, no one, unfortunately, can in particular be regarded as a specific. Moreover, some drugs recommended by some practitioners are not advocated by others. In all cases professional aid is necessary. A moderate dose of some purgative should be given in the first place, and the bowels afterwards kept open by a laxative diet of mashes and oatmeal gruel. Three or four drachms of aloes, or two drachms of calomel, may be given in the first place. Chloral hydrate, in doses of two to four drachms, may be administered three times daily in the water. It is not advisable to give drenches, as these annoy the animal in most instances; but remedies should be given in the water, or by clysters in the form of powders, or, lastly, by injection under the skin. In the latter method, morphia may be administered. Tobacco was found very useful in tetanus by the late Mr. D. Gresswell, and Mr. Charles Gresswell, of Nottingham, also advocates its administration. We also very strongly recommend it as the most valuable of all remedies in lock-jaw. The spine may be rubbed with the compound liniment of belladonna three times daily. It is very important in tetanus that the attendant be as quiet and kind to his charge as possible. He should always keep oatmeal or linseed gruel by the animal, and if it be impossible to take in sufficient nutriment in this manner, nutrient clysters are necessary.

RHEUMATISM.

RHEUMATISM assumes three different forms—acute, chronic, and muscular. Acute rheumatism is a constitutional fever characterised by special tendency to inflammation of certain parts, viz., the joints, the coverings of the muscles and of the "tendons" or "leaders," as they are sometimes called, and finally of the serous covering of the heart, and of its inner lining membrane. These inflammations have, as in man, a remarkable tendency to dissapear suddenly from one part and to reappear in another, without any apparent cause whatever.

Before speaking of the symptoms by which we may recognise the acute form of this malady, we may say a few words regarding the nature and the causes of all the varieties of rheumatism generally. Rheumatism is a general disease, the immediate cause of which is said to be some poisonous substance circulating in the blood. This poison is believed to be an acid. No acid, has, however, been detected in the blood. It seems not at all unlikely that rheumatism will eventually prove like so many other diseases of which we have already treated—to be due to some living germ or fungus circulating in the system.

The exciting causes of rheumatism are exposure to cold and wet, exposure to sudden chills, damp, and general bad hygienic conditions. It has been observed, as in man, that certain animals of the equine tribe are more pre-disposed to this malady than others, owing to a constitutional tendency or "rheumatic diathesis," as it is termed in medical language. Rheumatism is more common in some districts than in others, and is more prone to attack young animals than old ones ; moreover, those attacked are unfortunately more liable to future second, third, fourth, or even more frequent affections. Acute rheumatism very frequently causes inflammation of the valves of the heart, and in this is its chief danger ; for although the animal may entirely recover to all intents and purposes, nevertheless the heart may be left diseased, and the animal thus be unfit for prolonged exertion. Therefore, when a horse has had rheumatism, his value tends to be diminished for these very reasons, viz., his greater liability to future attacks and the probable permanent damage done to his central organ of circulation ; but it is not in every case that the heart is affected, and sometimes it may quite recover its normal condition.

We will now turn to the consideration of the symptoms of the various forms of this disease. In acute rheumatism there is high fever, the pulse is accelerated, and, if felt, will be found to be firm and full. The temperature is raised several degrees ; it may reach as high as 104° or even 106° F., and in this latter case is of very serious omen. The bowels are constipated, and the water passed is highly coloured, scanty, and generally clear and acid. In normal health it is cloudy and of an alkaline reaction in the horse, as in other herbivorous animals. There is sudden and severe lameness, with or without swelling of one or more joints, most commonly the stifle and fetlock, less generally the hock and knee. On manipulation the affected joints are found to be very tender. In most instances the heart is affected, but if the attack be slight, the symptoms of heart mischief may pass unobserved.

If, however, this organ be much affected, the animal will most likely exhibit pain on his left side over the region of the heart, which is very tender on pressure. The ordinary sounds caused by the heart's action will be altered in character, or replaced by what are termed "murmurs." In health, the heart while beating makes two sounds for each beat, and these have been compared to the words "lab, dup ;" but their absence or replacement by "murmurs," which sound like the letters " sh," cannot be recognised except by the initiated. The pain and swelling of the joints often subside in one extremity, and reappear in another, and this changing is a marked feature of rheumatism. Sometimes, but rarely, the inflammation does not abate, but proceeds, and "matter" or "pus" is formed in the joint or joints. When death does occur in acute rheumatism, it is nearly always due to disease of the valves or of the outer lining membrane of the heart. Now, regarding the treatment of this disease ; the animal should be warmly clad and placed in a well-ventilated box. A mild aperient, such as three drachms of aloes may be given in the first instance, and the bowels should afterwards be kept open by regulation of the diet, or if necessary by repetition of the purgative. Three or four times daily half an ounce of bicarbonate of potassium with an ounce of nitric ether may be given. If the temperature exceeds 103°, salicylate of sodium must be given in addition two or three times daily for two or three days, until the fever be reduced. The dose of this valuable remedy for the horse is four drachms. Locally, hot fomentations to the joints and anodyne lotions, as liniment of belladonna alone or mixed with an

D

equal quantity of liniment of aconite, will be found valuable. In the later stages, if the joint affections show little improvement, stimulating liniments of ammonia and camphor with turpentine may be applied, and in some cases even blisters may be required to reduce the inflammation.

Chronic rheumatism may follow the acute variety, or it may occur as an independent affection. It is not usually attended by much fever, and the inflammation has less tendency to shift from one place to another than in the preceding affection. The inflammation of the joints has a more lasting character, and more frequently leads to ulceration of the ends of the bones, on which excrescences may form. The joints may or may not be enlarged, and in some instances they may become fixed or anchylosed. When this disease is chronic from the first, laxatives should be given occasionally, and bicarbonate of potassium in half ounce doses, with one or two drachms of iodide of potassium may be given three times daily in the drinking water or as a draught. Tonics, such as iron with cinchona bark or nux vomica, are also required ; but this treatment as a rule is not very successful in chronic rheumatism. Locally, stimulating liniments are useful, and sometimes still more active treatment, as the application of the firing iron is necessary.

Muscular rheumatism is a very painful form, generally due to cold, damp, and fatigue. It mostly affects the muscles of the loins and buttocks, but may also involve the muscles of the neck, chest, and shoulders. Sometimes the affection is attended with slight fever. The back is elevated, and the affected muscles are tender and painful. The same remedies may be given as in the acute form, but for the salicylate of sodium, unless the fever be high, we may substitute two drachms of bromide of potassium. Locally, belladonna liniment is very useful in assuaging the pain, and hot fomentations also prove valuable. The animal should be kept warm in a loose, well-ventilated, but not draughty box.

RABIES, OR HYDROPHOBIA.

RABIES, or hydrophobia, is fortunately a somewhat rare disease in the horse. This malady is an effective febrile disorder originating in the canine and less frequently in the feline tribe, and occurring in the horse as the result of a bite of some rabid animal. It has been shown by M. Pasteur, to be due to a living vegetable germ or fungus, and the "virus," as it is termed, is transmitted through the saliva of the rabid creature. The disease itself is characterised by pain in the part bitten, great excitement, irritability, a disposition to bite, spasmodic seizures of the muscles prostration, and death. The disease generally manifests itself in the horse in from fourteen to forty days after being bitten, and it begins with great restlessness, excitability, and distress. The excitability increases, the animal becomes frantic, and attempts in his fury to destroy everything, and in some instances he bites savagely at the seat of injury. Febrile symptoms are also present. A flash of light or sudden noise, a disturbance of any kind, will bring on a paroxysm of fury. He has difficulty in swallowing, a characteristic hoarse cough, abundant flow of saliva from the mouth ; and the fits of violence are more aggravated and prolonged than they are in the dog. Gradually the fury becomes permanent, and the horse, prostrate, dies in convulsions on the second, third, or fourth day.

When the disease has set in, treatment is of no avail, and it is best to shoot the animal.

Bites by rabid animals should be treated at once. If possible, the tissues around the injury should be excised. If the wound be superficial, the application of caustic will be sufficient. If it be deep, the parts must be excised and then cauterised, or treated with caustics, such as nitrate of silver, carbolic acid, or caustic potash.

Recently M. Pasteur has devised a method by which he inoculates with what is termed "vaccine" or "attenuated virus," animals bitten by rabid creatures ; and by this means he claims to prevent the development of this dread malady. M. Pasteur has done so much in the way of practical preventive therapeutics that we have good reason to expect that this method, too, may prove as practically successful as his former wonderful and ingenious discoveries, of which his countrymen have indeed good reason to be proud.

Since writing the above for the columns of "The Yorkshire Weekly Post," more proofs of the value of M. Pasteur's treatment have been forthcoming.

CHAPTER II

DISEASES OF THE BREATHING AND CIRCULATORY ORGANS.

Coughing, Grunting, Whistling, Roaring. Asthma and Broken-Wind. Chill, Common Cold or Catarrh, and Chronic Catarrh. Sore Throat or Laryngitis. Bronchitis—Acute, Chronic, and Mechanical. Congestion of the Lungs. Inflammation of the Lungs. Pleurisy. Palpitation of the Heart, and Intermittence of the Pulse.

COUGHING, GRUNTING, AND WHISTLING.

HAVING now concluded our sketch of the general diseases of the horse, which are included under "medicine," we propose to treat in order of the several disorders of the breathing mechanism to which this animal is subject. Before commencing our description of the diseases of the organs, we must briefly refer to certain important symptoms associated for the most part with disorders of the respiratory tubes, and we may conveniently speak in the first place of "coughing."

Coughing is a symptom of various diseases, the signs and treatment of which will be described in their respective order. It is a modification of breathing, and it consists of a deep-drawn inspiration, followed by closure of the orifice of the main air tube at its opening into the back part of the mouth, and by one or more short but violent expiratory efforts. Generally it is excited by irritation at this opening, or in the breathing tube or its ramifications ; but sometimes it may be a nervous affection.

Cough is dry or moist. Dry cough is of several varieties,—short, hollow, hacking, broken-winded, and spasmodic. It is characteristic of irritation and of dryness of the lining membrane of the breathing tubes. In the early stages of inflammation it is loud, long, and sonorous ; and becomes rasping, and afterwards moist. In chronic disease of the larynx, or upper part of the air tube, it is loud, soft, and hollow. In the early stages of bronchitis it has a hollow metallic sound, and afterwards becomes moist, and is more or less painful throughout the disease. In acute inflammation of the lungs the cough is short, and in the later stages of the disease it is accompanied by expectoration of a rusty coloured secretion. In pleurisy the cough is dry and hacking, and is sometimes broken, as it were, in the middle.

The broken-winded cough is at first spasmodic, but becomes as the disease advances, feeble, short, and single. The animal being unable to relieve himself by the action of the chest and lungs, gives a suppressed cough, which is very characteristic and suggestive, even to the uninitiated. The hollow cough varies in degree, and indicates chronic mischief. Moist cough is indicative of an inflamed and humid condition of the lining membrane of the respiratory tract.

There is a cough spoken of as the "teething cough" of young horses. It is dry, and though more or less continuous, is of a more distressing character in the morning than at other times of the day. The age of the animal and the inflamed condition of the gums give us aid in detecting the nature and cause of this complaint, which is not due to cold, but is of a nervous nature, and not at all uncommon in four-year-old animals. The complaint is best combated by allowing only soft food for several days, and a mild oleaginous aperient, such as half a pint of linseed or castor oil every day for a few days. Half an ounce of bicarbonate of potassium, with two drachms of nitre given in the drinking water twice daily, will prove beneficial. Sometimes a horse may cough owing to irritation caused by indigestion or worms. The treatment of these complaints will be specified in due course.

Again, sometimes horses have a tendency to cough from a slight chill. There are all degrees of severity of such a cough; but when not due to active inflammation of the respiratory tubes or lungs, or to commencing influenza or other disease, it may be treated by allowing soft diet and the administration for several days of the medicines above mentioned. In case these remedies do not alleviate the complaint, the throat may be rubbed with compound liniment of camphor, or liniment of belladonna, or simple liniment of ammonia and turpentine, but the first is the most efficacious. A ball, also, containing one drachm of camphor, one drachm of ipecacuanha, one drachm of carbonate of ammonium, made up with gentian and treacle to one ounce, may also be given twice daily. Lastly, we may refer to chronic cough. This variety is almost always the result of bronchitis, influenza, or strangles, or it is the chief symptom of chronic bronchitis, or it may be left as a sequel after all but complete recovery from these complaints. It is a hard dry cough, and not at all uncommon, more especially among fast working horses, and often proves very inveterate. Soft laxative diet, of which green food and carrots should form main items, is very beneficial. Internally the above ball administered twice daily, is sometimes sufficient to cure the cough in a week or two ; but if not effectual the following formula may be substituted :—of camphor one drachm, of ammonium carbonate one drachm and a half, of iodide of potassium one drachm, of extract of belladonna one drachm, of gentian and treacle sufficient to make up to one ounce. We may here mention that no ball given to a horse should exceed nine drachms at the most in weight. A mild dose of physic should also be given occasionally. The work should be regular, but not too hard.

We shall shortly see that in some cases where heredity plays an activ part, chronic cough degenerates into roaring or broken-wind ; and for this very reason, if for no other, no horse affected with cough can during the continuance of the complaint be passed as sound.

We will now turn to the consideration of grunting. If a horse when struck or suddenly moved, makes during expiration a grunting sound he is termed a grunter. The emission of this noise is always to be regarded with suspicion, as it generally accompanies whistling and roaring. It may or may not depend upon diseases of the upper part of the breathing tube. In some cases a horse may grunt from pain alone, when suffering from pleurisy or from a neuralgic affection of the respiratory muscles of the chest, called pleurodynia, and other diseases. Many cart horses and large horses of any breed are apt to grunt, being nevertheless perfectly sound in their wind ; and, indeed, if fed for a time on heavy bulky food, any horse may become a grunter from this cause alone. If a grunter stands the tests used to detect roaring without making any noise in his breathing, he is, according to Professor Williams, and in the writers' opinion also, to be considered as sound.

Whistling is of two varieties, soft or moist, and dry or hard. The former occurs in acute inflammation of the larynx, when much exudation is thrown out in that structure, and also when the lining membrane is much swelled. In the first condition it is a wheezing noise, and is mostly diminished when the horse coughs. In the second case it is louder during inspiration than during expiration.

Soft whistling constitutes temporary unsoundness. It is in many cases unsafe to hazard a decided opinion for some days or even weeks, until the thickening of the lining membrane and the relaxed condition of the vocal cords have had time to regain their normal state. Dry whistling is, according to some authorities, to be regarded as a modification of roaring. Others, however (among whom is the writer), are of opinion that whistling and roaring are due to different states of the throat, and that they may exist independently of each other. Dry whistling, like roaring, is a sound made during inspiration. It is . due to diminished calibre of the larynx, or sometimes of its continuation downwards, owing to thickening of the lining membrane, distortion of the neck through tight-reigning, the presence of a fixed tumour in the air tube, or any other cause which diminishes the size of the passage through which the air escapes to and from the lungs. Whistling, though loudest in inspiration, is by no means absent during expiration. Dry whistling, like roaring, is often traceable to hereditary influence, and it constitutes unsoundness.

ROARING.

ALTHOUGH few complaints of the horse are so well known as roaring, yet there are not many regarding the nature and cause of which more erroneous notions are generally prevalent. Roaring is a loud unnatural sound made during inspiration. It is much louder than whistling, but is not of such a shrill character. In most instances, and of these we shall first speak—roaring is a chronic disease due to wasting and consequent paralysis of certain muscles of the larynx, or upper part of the windpipe ; but it may also be one of the signs of active inflammation of the larynx itself, in which case it is merely temporary, and does not constitute permanent unsoundness.

The origin of the wasting of these particular muscles is not certainly known, but we may mention that it more usually attacks those of the left side only, though it may sometimes involve those of the right side, and that it frequently ends in the paralysis of the muscles themselves. Now, as the muscles which open the larynx are those which become paralysed and unable to act, when the horse takes an inspiration, the characteristic sound called roaring is emitted. Roaring is in most instances gradually developed. At first the sound may be intermittent, and even weeks may elapse before it recurs after being once heard. As the muscles continue to waste, there is a corresponding and permanent loss of power or paralysis, and what at first was intermittent becomes established. In most instances, however, the noise is not intermittent, but gradually increases in proportion to the waste and paralysis of the muscles. The roarer generally emits a very characteristic cough in addition to the abnormal sound, and this cough is loud, deep, harsh, and dry. Most roarers, moreover, are liable to grunt in addition.

Hereditary influence plays a very prominent part in the transmission of roaring, and Professor Williams mentions a breed of horses in which nearly all the animals of both sexes are roarers. Horses and geldings are, however, more likely to become roarers than mares, which are but rarely so afflicted. Small ponies are rarely, if ever, affected with roaring. It has been noticed that animals predisposed to roaring in most cases suffer from inflammation of the throat from very slight causes, and the disease usually manifests itself after several such attacks of sore throat and cold. The long-continued use of a tight bearing-rein may induce this complaint, by distorting the natural shape of the larynx. Roaring may also be due to tumours or other diseases of the nose, or to tumours in the chest cavity, or to injuries or distortions of the main air tube, or trachea, as it is termed, or finally to any distortion or narrowing of the larynx itself.

It is said that in India roaring is almost unknown among horses bred there, in spite of the fact that many of their imported sires have been confirmed roarers, and that this complaint is equally as rare among Arab as among Indian horses. Horses also at the Cape are said to be but rarely affected. In what direction are we to seek for the explanation of this comparative immunity enjoyed by horses in these particular localities? There can be but little doubt that the influence at work is more probably climatic than dietetic, though both factors may each play their part. There is good reason to suppose that hay given in too large amount, especially when containing a large proportion of rye grass, is an exciting cause of roaring. Cart-horses, moreover, fed on large quantities of dry, hard straw and chaff, are probably also rendered more prone to this wasting paralysis ;

and although we cannot explain in exact terms the reason of this fact, we may mention, in passing, that the nerves ending in the stomach, are branches of the *same* nerve from which the muscles of the larynx are supplied with motor power. We shall see, in treating of asthma and broken-wind, what marked influence food has in the causation of these diseases, as well as in the complaint now under consideration. One other cause of roaring, is confinement in badly-ventilated, close, stuffy stables.

In cases of roaring which are not very pronounced, the characteristic sound is usually not heard unless the animal be made to go at a fair pace. Roaring in many cases is at first intermittent, gradually afterwards becoming established as the muscles waste more. In such cases, in the first instance, the sound is often heard at the beginning of exercise, and passes off as the work is continued. If the horse be worse at the end of a canter than at starting, he may be regarded as a pronounced roarer in all cases, excepting in those where there is some inflammatory action of the throat, in which case it is most likely only a temporary phenomenon. In trying a horse for his wind, it is customary to place him against a wall and make a feint to strike him. If he grunts he is further examined ; if not, he is made to cough by compressing the throat, and if the cough enforced sounds healthy, he is passed. It is, however, best to have the horse galloped, and to let him finish his run as he is going uphill. In the case of a draught horse, the animal may be made to draw a load at a fair pace up an inclined plane, when, if he be a roarer, the characteristic sound will be made. All forms of roaring constitute unsoundness, yet, in some cases of recent inflammation of the throat or lungs, the animal should again be examined after a reasonable interval, before being finally rejected. Some horses, as is well known, are apt to make a noise when pulling hard at the bridle, owing to the pushing back of the tongue ; but this is easily remedied.

We will now consider the treatment of roaring. When the complaint follows influenza, strangles, or ordinary cold, the animal must be carefully treated and attended to. He should have light work only ; and a good nutritious diet, with not too much dry hay, should be supplied. The throat should be smartly blistered with a mixture of equal parts of ointment of cantharides and ointment of iodide of mercury, and a draught containing iodide of potassium one drachm, tincture of nux vomica one drachm, liquor arsenicalis two drachms, and water to a pint, may be given twice daily. In those cases of roaring which are dependent upon tumours, the latter should be removed when practicable.

The greatest number of cases of roaring are, as we have said above, due to actual waste of the muscles of the larynx, and these cases are the most inveterate. Blisters, or the application of the firing iron, have, however, succeeded in arresting the wasting of the muscles, when applied in the early stages to the skin in the region of the throat. Chlorate of potassium in doses of one or two drachms is recommended as well worthy of trial for arresting this wasting change, and with this view it may be given in the drinking water in the confirmed cases twice daily. If preferred, it may be given in doses of two drachms, with three drachms of liquor arsenicalis, in the form of a draught twice daily with a pint of water, after feeding. In very bad cases, the sound may be lessened by pads attached to and fitted carefully over the nostrils, so as to regulate the amount of air entering the larynx. If this method is effectual, a tube may be passed through an opening made in the windpipe, and kept there for the remainder of the life of the animal. The electrical current, together with the use of such drugs as nux vomica, the iodide, the chlorate, and bicarbonate of potassium, liquor arsenicalis, and arseniate of iron, constitute probably the most useful of all combined methods of treating roaring. In Germany, the removal of one of the cartilages of the larynx is recommended; but this treatment is not very successful, and we therefore pass it by. Bad roarers can be used for slow work, but they thrive badly as a rule, and often succumb to slight disorders, more especially of the breathing organs.

ASTHMA AND BROKEN-WIND.

ASTHMA of the horse is a morbid condition, characterised by attacks of difficulty of breathing. It probably depends as in man, upon spasm of the small air tubes in the lungs, and is often accompanied by a wheezing noise, which is more distinct than in the allied disease.

One of the chief predisposing causes of asthma is inherited tendency; but fatigue, overwork, general debility, and other factors also play a prominent part in its production. It bears a close resemblance in many points to broken wind, of which we shall treat shortly; and if prolonged, it not unfrequently terminates in this chronic malady. By some authors asthma is regarded as an early stage of broken-wind, but as many cases undoubtedly recover without ever passing into this more serious condition, and as the treatment required is essentially different, we have thought well to consider it separately. Asthma is characterised by sudden spasmodic difficulty of breathing, which resembles that of broken-wind, in that the inspiration is easier than the expiration. The latter is usually of a jerky character, but has a less distinct double action than in the allied malady. In asthma also the wheezing noise made is more distinct, and there is more exhaustion with less cough, which is not so hollow as in broken-wind, but is short, quick, or suppressed. The suddenness of attacks of difficulty of breathing, their severity, their rapid accession and decline, and their unaccountable disappearance, are marked features of asthma. The febrile disturbance is severe when the attack is fully developed, the chest is fixed as it were, and there is increased movement of the muscles of the belly. When the cough is severe, small pellets of mucus are discharged through the nostrils. If we listen to the fore part or to the side of the chest, we can distinctly hear the wheezing sound. The spasms may last a few days, or may extend over several weeks, and then disappear or pass impercept-ibly into broken-wind. In attacks of asthma, the horse should be placed in a well-ventilated roomy box, and the diet should consist of bran mashes, and oatmeal or linseed gruel. If the cough be severe, the sides of the chest may be blistered with equal parts of cantharides ointment and ointment of red iodide of mercury. Half an ounce of bicarbonate of potassium may be given in the water twice daily. During the acute stages, a draught consisting of bromide of potassium two drachms, spirit of chloroform one ounce and a half, spirit of nitric ether an ounce, water to a pint, may be given every six hours, or every four hours if necessary. Should a very severe paroxysm occur, the horse may be made to inhale forty minims of nitrite of amyl poured on a sponge held over the nostrils. The attendant should be careful not to breathe this vapour. We will now consider the chief points connected with broken-wind.

Broken-wind is characterised by difficult and spasmodic breathing, the inspiration being easily performed, the expiration being very prolonged, and accomplished by two apparent efforts. The difficulty of breathing is constant, and though marked by exacerbations, and by periods of greater ease at times, it is not truly intermittent, as in asthma ; and the cough, spoken of as the broken-wind cough, is short and nervous. Indigestion, flatulence, and heart disease, sometimes aggravate the difficulty of breathing.

The exact nature of broken-wind is still a disputed point. According to Professor Gamgee—and with his view on this point we are entirely agreed—broken-wind is at first a purely nervous affection depending on an unhealthy condition of the organs of digestion, and the changes we find in the lungs are due to such nervous disturbance. The condition of the digestive organs is to be attributed to improper dieting, or to constitutional predisposition to digestive troubles. Around the small breathing tubes of animals are layers of muscle fibres, and when these latter contract, they aid the expulsion of the air from the lungs. In broken-wind they are first spasmodically contracted, thus interfering with the passage of the air, and causing the difficulty of breathing, and then they afterwards become paralysed, and finally undergo decay. This is owing to the irritation set up by indigestible food, acting on the branches of nerves supplying the stomach. The disturbance thus arising is then reflected to the breathing organs. This spasmodic contraction and paralysis, while it interferes seriously with both respiratory acts, chiefly obstructs the expiration. The expiratory action thus becomes double, since a double contraction of the muscles of the belly is required in order to force out the air from the diseased air cells and small air tubes of the lungs.

Round-chested horses sometimes become broken-winded without any apparent cause, and difficulty on expiration in such cases may be attributed to the limitation of the movement of the chest. It will easily be seen that when the chest is round, the movements are more limited, and the horse will in consequence not be able to take deep inspirations or make strong expiratory efforts. Other changes in addition to those above spoken of now take place in broken-wind, in consequence of the paralysis of the layers of muscle fibres surrounding and forming one of the walls of the small air tubes. The little air-cells of the lungs become inflated with air, and the nutrition of their walls becomes interfered with. Wasting of the walls now follows, and air accumulates among the tissue which binds the air-cells together. The heart, now having more work to do in consequence of these changes in the lungs, becomes accordingly enlarged on the right side. Now that we have seen what the proximate causes of broken-wind are, let us review its remote causes. The first is hereditary influence. The offspring may inherit the same bodily conformation and temperament as the parents, and thus be liable to be similarly affected. Again, defective dietetic conditions are largely responsible for the production of broken-wind. As the diet of the coarser breeds is frequently innutritious and bulky, and the animals are worked after a heavy meal with full allowance of water, they are more frequently affected with the disease than better bred horses which are more carefully attended to. Chopped hay and oat straw in large amounts are also said to be potent causes of broken-wind. Finally, sometimes the malady may follow as the result of previous inflammation of the lungs.

The symptoms in confirmed cases of broken-wind are unmistakable, but when not so fully developed, the disease may be occasionally overlooked. The inspiratory movement is performed rather quickly and with ease, while the expiratory act is more prolonged, difficult, and accomplished by two apparent efforts. It begins rapidly, and is suddenly stayed before the act is finished. The cough is characteristic, being short, of little force, and suppressed. It seems to be ejaculated with a kind of a grunt through the upper part of the windpipe. When the animal is in fair condition, the cough generally occurs only at long intervals, and rarely in paroxysms ; but at the commencement of the disease, and when the horse is excited from any cause during exercise, or at any other time, it is apt to be very severe and continued. In many instances cough is one of the first 'indications of broken-wind. After feeding, the symptoms are more severe, and they are liable to exacerbations from extremes of heat and cold or other atmospheric changes. The chest being rounder than it should be, its movements are much impaired ; while the movements of the belly are violently put into action during an attack of coughing. In well marked cases, a loud sonorous wheezing noise can be distinctly heard by those near the animal. In confirmed cases the digestive organs are weak and easily deranged. The horse is debilitated, unthrifty, and the coat is often harsh, dry, and scurfy. When worked he is easily fatigued, and perspires readily ; and the bowels are generally loose.

Finally, we turn to the treatment of this disease. Above all things it is essential that the diet should be carefully regulated. The food should be nutritious, digestible, and in moderate quantity, and the water supply should be well regulated. Dry hay should not be allowed, but freshly mown grass or lucerne and carrots may be substituted with great advantage. The corn may be bruised and damped. The animal must never be worked immediately after a full meal. The general hygienic arrangements should be looked after, the stable should be well ventilated, and the general health attended to. The symptoms of broken-wind may be ameliorated by all remedies which improve the general condition and the digestive powers of the animal. Liquor arsenicalis may be given in the drinking water, in half-ounce to six drachm doses twice daily for two or three weeks, then once a day for a similar period, and finally once every alternate day. Purgatives, such as aloes or linseed oil, may be given in moderate doses if required. Their occasional administration is beneficial in regulating the action of the bowels. It is well known that horse-coopers adopt certain measures in order to pass a broken-winded animal for sale. With this object they allow little or no food, give the animal a good sharp trot to empty the bowels, and administer drugs such as digitalis, opium, and other agents such as shot, which have a temporary sedative effect, and thus deceive many persons.

CHILL, COMMON COLD, AND CHRONIC CATARRH.

IT not unfrequently happens after a day's hunting, more especially when there have been many and prolonged halts on cold and wet days, that horses take a chill in consequence of the exposure. Frequently the services of a veterinary surgeon are not called for in these cases of simple fever, or febricula as it is termed, unless the symptoms be somewhat more severe or the animal more distressed than he generally is. In most instances of simple fever resulting from chill, the attendant first notices that the animal does not take his food, but stands dejected or moves about restlessly, with cold and staring coat. The number of respirations is not much increased in ordinary cases, but may reach as high as 18 or 20 in the minute. If much more frequent, we have reason to suspect that the lungs may be inflamed or congested, and the case is then, of course, of a much more serious nature. The pulse in simple fever is raised from 48—60, or even possibly higher, and the internal temperature as indicated by the thermometer reaches to 102°—104° F. The bowels are constipated, and the fever remains high for two, three, or four days, or possibly longer. In such cases as these, it will be necessary in the first place to put the animal in a well-ventilated loose box, and the diet should be of a laxative character, consisting of linseed or oatmeal gruel, a few carrots, and grass. The groom should carefully bandage the legs, and should keep up the surface heat by friction of the legs and ears, and by moderate clothing. Medicinally, either half a pint of linseed oil, or two or three drachms of aloes should be given in the first instance: and a draught consisting of Fleming's tincture of aconite, five minims ; liquor ammonii acetatis, four ounces ; bicarbonate of potassium, half an ounce ; nitric ether one ounce and a half ; and water to make a pint, may be administered every five or six hours, so long as the acute symptoms last. Such cases, however, generally recover by the end of the second or third day, when they are not complicated by any other malady.

We have now to speak of acute catarrh or common cold, as this complaint is usually termed. By the term catarrh, we mean a condition characterised by inflammation of the lining membrane of the nostrils, and of its continuation along the upper portions of the windpipe. In this condition there is a discharge from the nose, and occasionally cough and sore throat. Symptoms of fever are also sometimes present. The causes of catarrh are sudden variations in the temperature, exposure to cold and damp, hot and badly-ventilated stables, and contact with affected animals. Young animals, when first brought up into warm stables, are especially liable to attack. During the change of the coat there is also great predisposition to catch cold. The symptoms are sneezing, redness and dryness of the membrane lining the nostrils, followed by a discharge, which is at first thin, but soon becomes turbid, yellowish-white, and profuse. There is also redness of the membrane lining the eyelids, with discharge of tears and drooping of the head. Febrile symptoms are sometimes manifested, and vary much in intensity. The temperature may rise about three degrees, or even a little higher. The pulse and respiration are also usually accelerated, and the appetite is impaired. Debility and general dulness frequently supervene. Such cases as these almost invariably terminate in recovery. In mild cases, rest from work in a warm and well-ventilated but not draughty loose box, with attention to the diet will suffice. In all cases where the febrile manifestations are at all severe, a febrifuge draught, consisting of liquor ammonii acetatis four ounces, of nitric ether one ounce and a half, of chlorate of potassium one drachm and a half, and sufficient water to make a pint, may be given three times daily. When the bowels are much confined, enemas of warm water may be given, and, if necessary, two or three drachms of aloes may be administered in addition. If the throat be sore, and the cough troublesome, we may administer in addition one drachm of camphor. In the early stages, while the membranes are dry, inhalation of hot water vapour is useful in relieving the irritation. When the throat symptoms are severe, compound liniment of camphor, or liniment of turpentine should be applied externally. The diet should be laxative, consisting of scalded oats, oatmeal, or linseed gruel and green food.

Chronic catarrh of the nose, or chronic nasal catarrh, is a discharge of varying character from the nostrils, and it may be continuous or irregular. Most of these cases are due to an unhealthy condition of the membrane lining the nose, and are the result of protracted and severe acute catarrh. It may also arise from external injuries, decay of the upper grinders, and other conditions. There is a discharge of a greenish, purulent fluid, and the membrane lining the nose is of a leaden hue, or it may be blanched and thickened. The general health is generally somewhat impaired. In these cases the animal should be rested, and liberal diet allowed. Internally, a draught containing two drachms of citrate of iron and ammonium, and two drachms of carbonate of ammonium, with one drachm and a half of tincture of nux vomica, may be given in a pint of water twice daily. Locally, lotions consisting of four to twenty grains of sulpho-carbolate of zinc, in each ounce of water, may be injected up the nostrils by means of an enema syringe or through a nasal funnel. In many cases the insufflation of atomised solids is to be preferred to lotions. Equal parts of iodoform and starch finely powdered, blown up into the nostrils in quantities of about a drachm at a time, will be found a very efficacious remedy.

SORE THROAT, OR LARYNGITIS.

WE have incidentally, in treating of the several fevers and of common cold, alluded to sore throat or inflammation of the upper portion of the windpipe, or larynx, as this part is termed, and we have occasionally spoken of bronchitis. We have now to consider these inflammations separately, their varities, causes, symptoms, and treatment.

There are two chief forms of laryngitis or sore throat in the horse, the acute catarrhal and the œdematous variety. The first is the simpler form, and is dependent upon the same causes as common catarrh or cold, of which, indeed, it is usually one of the earliest and most prominent symptoms. There is pain and difficulty in swallowing, and the throat shows signs of great tenderness when handled. Usually there is cough, at first hard and sonorous, but afterwards becoming less resonant. Between the branches of the lower jaw there is swelling, and this may sometimes occur over the side of the face also. Discharge from the nose may or may not occur, although it is a constant symptom when catarrh is also present. Symptoms of fever are more pronounced than they are in simple catarrh. The appetite is diminished or lost, the temperature rises, and the pulse is accelerated. In the second or œdematous form, the symptoms are more aggravated and dangerous. This disease may succeed what at first appeared as an ordinary case of the catarrhal form, or it may begin suddenly and run a very rapid course. It has also been met with as the result of inhalation of poisonous acrid vapours and hot air. In these cases sometimes the swelling and the effusion into the structures of the throat become very considerable, and the breathing very much disturbed. The pulse is quickened, the temperature rises, and the membrane lining the nose becomes of a purplish hue, owing to deficient aeration of the blood. The nose is protruded, the upper air passages being thus made to approach as near as possible a horizontal line. When the swelling of the inner laryngeal structures becomes very great, the respiration is suddenly difficult, and the inspiratory action is especially prolonged, and accompanied with a peculiar harsh sound, succeeded by a short expiration. The nostrils are then dilated to their full extent, the face has an anxious expression, and there is great distress. The extremities are cold, and sweats bedew the body ; the animal stamps with his feet, and his distress still increases ; the visible membranes, such as that lining the nose become more livid, prostration ensues, and the horse, unless relieved, soon succumbs. A short time ago, the writer had under his charge, a valuable hunter with this severe form of sore throat, and the animal would have soon succumbed, had vigorous measures not been forthwith adopted.

In severe cases of sore throat, the treatment should be as prompt as possible. The animal should be placed in a large airy loose box, and in no case confined in a small stuffy stable, as his chances of recovery will be greatly lessened unless he have a good supply of pure, fresh air. Inhalations of hot water vapour, medicated by the addition of carbolic acid, are of great value in all forms of acute sore throat, and often greatly relieve the difficulty of breathing. Hot fomentations to the throat are also very useful; but if the general distress and difficulty in breathing continue, it may be necessary for the veterinary surgeon to open the windpipe, in order to avert suffocation. This extreme step, however, is not often required, The hot water vapour may be generated by pouring hot water over chopped hay or bran in a nose bag, and may be medicated by the addition of one ounce of tincture of opium to the quart of water, or as above. In the mild forms of larnygitis, the inhalations and fomentations to the throat may be followed up by the application of stimulating liniments to the outside of the throat.

Internally, in sore throat, febrifuges may be given, in the form of draughts administered every four hours. The formula prescribed in common cold, will prove likewise very serviceable in this disease. The diet should be soft and laxative, and the animal should be warmly clad and carefully attended to. When swallowing is very difficult, it will be necessary to give all medicines in the water or food. Belladonna is a useful drug in the early stages of sore throat, in addition to the above remedies; and two drachms of the extract may be administered twice daily, by placing the medicine between the horse's teeth in the form of an electuary. During recovery, the diet should be as nutritious as possible, and vegetable and mineral tonics should be given, in order to combat the great prostration usually left in these cases. The formula for a tonic mentioned in treating influenza would prove very serviceable in this disease also.

BRONCHITIS, OR INFLAMMATION OF THE BRONCHIAL TUBES.

BRONCHITIS is an inflammation of the lining membrane of the bronchial tubes or prolongations from the windpipe into the lungs. Sometimes it is limited to the large tubes, or it may extend to the ultimate ramifications. The causes of bronchitis are debility; previous attacks of bronchial inflammation; exposure to cold and damp; irritation of the tubes by noxious vapours, or by the accidental entrance of fluids or solids into the bronchial tube. Bronchitis also often accompanies influenza, and is met with in certain other fevers, and under various malhygienic conditions. We thus see that there are three forms of bronchitis—*primary; secondary, i.e.,* coexisting with, or following after, certain fevers and other diseases; and *mechanical,* or depending on noxious vapours, fluids, or solids, irritating the lining membrane of the tubes. Acute bronchitis is usually ushered in with chilliness, malaise, and febrile symptoms, though in many cases these may be trivial. When fully developed, besides the symptoms of ordinary catarrh, there is a frequent hard and sonorous cough, which gradually becomes of a softer kind. The appetite is impaired, and the horse is dull and dejected. The pulse is increased in number and is rather soft. The respirations are much accelerated, being relatively much higher than the number of pulse beats, and in many severe cases they are as numerous as the pulse, and may exceed it in number. As the disease progresses, there is expectoration of a scanty, ropy, tenacious mucus. The discharge escapes to some extent through the nose, but the greater part passes into the mouth and is swallowed. In the later stages, however, a profuse discharge escapes through the nostrils. The cough becomes more violent and frequent than at first, but gradually becomes less severe, and finally disappears. Bronchitis of the larger tubes, ends in most instances in perfect recovery ; but when affecting the small tubes and vesicles of the lung, it is always dangerous, and requires great care. Mechanical bronchitis is induced by the inhalation of some irritant. The irritating agent may be gaseous, as for instance smoke from a burning building or acrid fumes ; or it may be fluid or solid, as for instance water, food, or other matters which perchance find their way into the air tubes.

In cases of bronchitis, the horse should be warmly clad and placed in a well-ventilated loose box, and the diet should be liberal, nutritious, and laxative. The animal should be made to inhale hot water vapour, which may be medicated by the addition of two drachms of carbolic acid to each quart of hot water. If there be any sore throat, stimulating embrocations, as for instance compound liniment of camphor, should be applied externally. If the bowels are inactive, enemas should be administered, and if they still continue constipated, three or four drachms of aloes, or three quarters of a pint of linseed oil may be given. A draught made up of four ounces of solution of acetate of ammonium, one ounce of nitric ether, one drachm of camphor, and half an ounce of tincture of squills may be given three times daily with half a pint of water. In cases where the small tubes are much affected, blisters or stimulating liniments may be applied with advantage to the sides of the chest. In the later stages tonic medicines are often required.

Chronic bronchitis is met with in the horse either as a sequel to the acute form, or as an independent disease. It differs from the acute form in its slower progress, and in its symptoms being less severe; and is characterised by a persistent hard and sonorous cough, and by the absence of febrile manifestations. In many instances, this disease causes gradual loss of flesh, diminution of appetite, and general debility. When it occurs as an independent affection, it is generally gradual in its onset and development, and of a very persistent nature when once established.

The diet in this complaint should be liberal and nutritious. Rest is not necessary, though severe exertion should be prohibited. The remedies recommended in the acute form may be with advantage tried in the more persistent and chronic form of the malady. A ball containing one drachm of camphor, one drachm of nitre, one drachm of ipecacuanha, and one drachm of squills may be made up to eight drachms, and given three times daily. Later on tonics are required. One drachm of powdered nux vomica, two drachms of carbonate of ammonium, and two drachms of citrate of iron and ammonium, made up into a ball of eight drachms with gentian and treacle, may then be given twice or three times daily. Chronic bronchitis is an exhausting disease, and unless carefully and judiciously managed, not uncommonly passes from bad to worse, until the animal loses its appetite, and becomes emaciated and incapable of work.

CONGESTION OF THE LUNGS.

CONGESTION of the lungs, or pulmonary congestion, is one of the most important diseases to which the horse is liable. It is therefore essential that our readers should have a clear and thorough knowledge of its distinguishing features, and of its mode of treatment ; for, indeed, not only is it one of the most preventible of maladies, but it is at the same time one which is in most instances thoroughly amenable to early, vigorous, and judicious treatment. Congestion of the lungs is met with in the horse, not only during the progress of many diseases, such as inflammation of the lungs, of the feet, and of the bowels, and in injuries of the joints, in heart disease, and in some contagious fevers, but also as a distinct and independent affection. It is the latter kind, which we propose to discuss ; for that form which complicates other diseases, is treated of, in connection with the primary malady. The kind of congestion which results in some forms of heart disease is called passive, and this may also be due to the general exhaustion resulting in some fevers and other debilitating conditions. The kind which occurs as an independent affection, is called acute or active congestion. It is the more frequent of the two forms in the horse, and is the more important and easily recognised. When it is accompanied by bleeding from the nose, from rupture of the small vessels in the lungs, it is sometimes spoken of as pulmonary apoplexy.

When an animal in an untrained condition is suddenly called upon to perform any unusual exertion, the heart, lungs, and muscles may not be able to respond to the increased strain put upon them. Under these circumstances —as, for instance, when an untrained horse is suddenly put into the hunting-field—the heart's action becomes embarrassed and tumultuous, the blood accumulates in the small vessels of the lungs, and the breathing becomes more and more distressed, until, at length, the horse may die of suffocation, consequent on over-loading and engorgement of the vessels of the lungs with impure blood. Under careful training, the heart and the other organs are gradually accustomed by regular and careful exercise to perform additional work. Their tone and vigour is enhanced, and the system responds duly to even severe strains.

The symptoms of the acute form of pulmonary congestion, are in most instances of a very severe type. The horse stands with his limbs out-stretched, and gasps for breath. All the muscles which can possibly aid in respiration are called into action ; the nostrils open and close in quick succession, and the flanks heave to and fro with great rapidity. Cold sweats bedew the surface of the body ; the extremities become very cold, and the lining membrane of the nostrils shows, by its livid hue, the condition of the blood circulating in it. The pulse is much quickened, and may reach 100 to 140 beats per minute. It is feeble, oppressed, indistinct, and becomes almost imperceptible in severe cases ; and there is a tremor all over the body. The heart's action, irregular and tumultous from the first, becomes still more embarrassed. The lungs become more engorged, and the breathing still more distressed, until, at length, unless treatment prove availing, death

E

results from suffocation. In more favourable instances, however, the engorgement subsides, the heart regains power, the circulation through the lungs is restored, and the animal soon regains its normal condition.

In some instances, frothy blood is discharged through the nostrils owing to rupture of the engorged vessels of the lungs.

Except in very severe cases, the animals usually make a complete and comparatively rapid recovery. Acute congestion of the lungs, however, is very liable to recur for some time after apparent recovery, and is not unfrequently followed by acute inflammation of the lungs. Instances of death from acute congestion of the lungs have not unfrequently occurred within a few days of the purchase of a horse, and the changes found after death, which in some cases are of a very marked character in the lungs, have sometimes been ascribed by the uninformed to long standing diseases of these organs. On some occasions, indeed, the seller has, in consequence of this mistaken idea, been compelled to refund the full value paid for the animal. Yet these very features, which were attributed to old standing disease, are on the contrary the characteristic results of acute congestion. Indeed, the darkly coloured friable condition of the lungs, with the tendency to putrefaction and liquidity, so far from being the results of chronic disease, are in all cases characteristic of acuteness of attack. No doubt, in many cases, the purchaser of the animals, presuming them to be in a well-trained condition, forthwith puts them on trial, with the result that congestion of the lungs sets in severely, and the horses die. The writer had under his care a very severe case of pulmonary apoplexy or congestion. The animal was a very fat draught mare. The attendant had taken her for an unusually long journey, and, on arriving home, observed blood oozing from the nostrils of the animal, which was standing with outstretched limbs panting and gasping for breath. She, however, made a complete recovery in a few days.

We will now turn to the consideration of the treatment of these cases. Of their prevention by careful training, and not putting animals to sudden unusual strains, we need not speak further here.

In the treatment of acute congestion, it is imperative above all other things that the horse should have a plentiful supply of pure, fresh air; and strict quietude and repose should be enjoined in all cases. The body must be well rubbed down with wisps of straw, and afterwards warmly clad. The legs should be carefully bandaged with thick wool or flannel, being previously rubbed with some stimulating liniment. Internally, alcohol, in its various forms, in moderate and frequently repeated doses, is of *great value* in helping to restore the flagging circulation. From four to six ounces of brandy, with three ounces of liquor ammonii acetatis may be given at first, every two hours for three times, and then every four hours for about the same number of times. If the symptoms continue unrelieved after these steps have been taken, bleeding is necessary in order to relieve the congestion of the lungs and the engorgement of the great veins and right side of the heart, a consequence of the impeded circulation in the lungs. By this means impending suffocation is averted. From three to four quarts of blood may be withdrawn by the operator. It is not advisable to repeat the bleeding, nor to apply mustard or other irritants to the sides, as these measures merely annoy the animal, and thus increase the difficulty of breathing. Cloths, however, wrung out from hot water may be applied closely to the chest. After being well wrung out from hot water and applied, a dry rug should be placed over them, the whole being fastened with a surcingle. These hot cloths should be renewed at intervals of one hour and a half, or two hours. In order to guard against inflammation of the lungs, which sometimes succeeds acute congestion, as well as to prevent a recurrence of the congestion itself, careful management is required for some time after the abatement of the acute symptoms. The diet should be light and nutritious, and water may be allowed freely from the first onset of the disease. The box in which the horse is placed should be well ventilated, but not draughty, and the body must be kept warm with clothing.

INFLAMMATION OF THE LUNGS.

INFLAMMATION of the lungs or pneumonia, rarely occurs alone in the horse, but is mostly associated with bronchitis and pleurisy. We have already mentioned that it not unfrequently follows congestion of the lungs. The usual exciting causes of this affection are sudden chills, exposure to wet and cold, especially after severe exertion or fatigue, and confinement in draughty or foul and badly-ventilated stables. It is especially prevalent during spring and autumn, when sudden changes in the atmospheric condition are of frequent occurrence. It may complicate specific fevers, such as influenza and anthrax.

Pneumonia is frequently ushered in by a severe shivering fit. The horse becomes dull and dejected, and the pulse, though variable, is generally accelerated, and often reaches 90—100 beats per minute. The breathing also is quickened and shallow, sometimes reaching as high as 50 or 60 per. minute, and if pleurisy be also present, it is painful, and though the chest walls move but little, the belly heaves quickly to and fro. The temperature is raised, and may be from 103° to 106° F. The skin and extremities are cold, and the membrane lining the nostrils is red and injected. The bowels are constipated, the horse loses his appetite, and wanders to and fro in his box in a dull, dejected manner, showing no inclination to lie down. There may be a dull, dry cough, not of that suppressed and painful character so noticeable in pleurisy. There is seldom much expectoration, though rusty or blood-stained, more or less viscid, tenacious matter is sometimes discharged through the nostrils; whereas in bronchitis it is more or less purulent and yellowish. As the disease progresses, the respirations, which at the outset are not much accelerated, become more rapid until the crisis, when they are, as mentioned above, much quickened and shallow. The breathing is also sometimes much accelerated in paroxysms, which are not infrequent during the progress of the malady. The febrile symptoms extend over a period of several days, or even longer. In favourable cases they then subside, and the cough, which becomes moister and more easy, gradually ceases.

Inflammation of the lungs is a dangerous affection, and requires careful treatment and management. The horse should be placed in a well ventilated, but not draughty loose box, the temperature of which should be kept at 70° to 75° F. The body should be clothed with rugs, and should be gently rubbed down occasionally with wisps of hay. Bleeding is necessary in the case of heavy draught horses kept in very high condition, as many of the agricultural and draught horses are, and also plethoric horses of other breeds, when the difficulty of breathing is very great and the fever high. In such cases it is our practice to remove from four to six quarts of blood from the jugular vein. We may allow the horse as much tepid or warm water as he will take, and with this object should leave a moderate supply by him. The diet should be laxative and nutritious, consisting of bran-mashes, linseed and oatmeal gruel, hay in moderate quantity, and roots or grass. After the subsidence of the fever, stronger and more nutritious diet should be substituted.

During the fever, a draught, containing four ounces of liquor ammonii acetatis, five minims of Fleming's tincture of aconite, one ounce of nitric ether, and two drachms of nitrate of potassium may be given with water to half a pint every four or five hours, until the acute symptoms abate. Active purgatives should not be given in this disease, but if there be great constipation, half a pint of linseed or castor oil may be given. In the later stages, and where there is marked debility, stimulants are required, and six or eight ounces of whiskey may be given three times daily, and may be persevered with if it prove beneficial.

During convalesence, tonics are required, and the formula mentioned in treating of influenza will be found useful. With regard to the local applications in pneumonia, when the extremities are very cold, they must be rubbed with some stimulating application of a non-irritating kind, and woollen cloths wrung out from hot water may be assiduously applied with great advantage around the chest. These should be renewed every two hours or oftener, as long as the disease continues in the acute stage. This hot pack should be closely applied, so that no cold air can pass between the rugs and the skin; and the temperature should be as high as the animal is able to bear. A good method of applying this treatment is to obtain a piece of felt about an inch thick and a foot and a half wide, fitted with straps. After being well wrung out from hot water and applied closely to the chest, a waterproof lined with flannel should be strapped round the felt. A simpler method is the application of an old blanket, wrung out thoroughly, and folded three or four times. Over it is placed a dry rug, the whole being fastened by a surcingle or line.

PLEURISY.

PLEURISY, or inflammation of the lining of the walls of the chest and the lungs, is frequently set up by exposure to cold and vicissitudes of temperature. We have already seen that it is frequently associated with, or supervenes after

inflammation of the lungs and bronchitis, and also that it is a frequent
concomitant of influenza. Sometimes pleurisy is ushered in by a slight chill,
at other times by a pronounced shivering fit. The animal is frequently
restless, and shows signs of pain, aggravated by moving or breathing, which,
though quickened, is performed carefully. The pulse is increased, especially
when the horse is moved round, and the chest wall is very tender. The
expression is anxious, and indicative of great pain. The pulse is increased
in frequency, and it is hard, and firmer than in the last affection. The
temperature is not so high as in the last disease, ranging from 103° to 104°.
The mouth is hot and dry, but the expired air is not so much heated as in
pneumonia. In the inspiratory act the ribs are fixed, and in consequence of
this, a furrow which is called the "pleuritic ridge" is formed. This extends
from the bottom of the back part of the chest, and runs obliquely in an
upward and backward direction to the hips. Cough is a frequent symptom
of pleurisy. It is shorter and more painful than in inflammation of the lungs,
and is attended with no expectoration. Not unfrequently during expiration
the horse gives a grunt, when he is moved. If the ear be applied to the
chest, a creaking sound may be heard. This is owing to the rubbing of the
dry and inflamed surfaces of the lining membrane together. This sound
appears as if close under the ear, and, as the lung moves to and fro, it is
consequently double. It is not heard if the breathing ceases for a moment.
The area over which it is audible may be very limited. This friction may
sometimes even be *felt* by placing the hand over the chest wall. The
disease now subsides, and fluid is poured out from the inflamed membrane
and accumulates in the chest. When this happens, and the effusion
increases, the symptoms become more severe. The pulse is raised to 80 beats
or more per minute, and is weakened and irregular. The breathing becomes
more laboured, and performed with very great difficulty. The flanks heave,
there is flapping of the nostrils, and the horse's head is generally protruded.
Dropsical swellings may appear in various parts.

Pleurisy generally involves only one side of the chest, in most cases the
right. Sometimes this malady has been mistaken by the uninitiated for
colic ; but the tenderness of one side, the constant pain, the high fever, the
friction sound, the altered breathing, and the "pleuritic ridge" seen between
the ribs and the belly, guide us in diagnosing pleurisy. The fever soon
subsides, and the pain diminishes in favourable cases, but where much
effusion is poured out, the disease often lasts several weeks, and may prove
fatal.

The treatment of this malady is very similar to that of inflammation of
the lungs, and therefore we need not repeat it. The woollen cloths should
be likewise carefully applied in pleurisy. If after the abatement of the
severe febrile symptoms, the animal seems to make no progress towards
recovery, but still breathes with difficulty, the temperature remaining high,
stimulating liniments may be applied to the chest, and four to six ounces of
whiskey may be given three times daily. If we have any reason to suspect
the accumulation of liquid in the chest, one drachm of iodide of potassium

may be given in the drinking water, two or three times daily. Where much debility follows the acute symptoms, tonics are required, and the formula mentioned in treating of influenza, will prove very useful here also. In some cases where fluid still remains unabsorbed in the chest, the veterinary surgeon passes a trocar between the ribs, and draws off the effused liquid, which may amount to many pints.

HEART DISEASES

WE shall not have much to say concerning diseases of the heart.

Palpitation or violent and tumultuous action of the heart, is a common symptom met with both in functional, and in organic disease of the heart. It is due to a variety of causes, such as sudden demand for work from the heart in excess of its powers. In cases dependent upon weakness, tonics are required, and the diet should be nutritious and digestible. When palpitation results from nervous excitement, quietude should be enjoined. Spirit of chloroform in one ounce doses, with five minims of Fleming's tincture of aconite, three times daily, is very useful. When we purchase a horse it is always well to feel the pulse, as intermittence or irregularity of the beat leads us to suspect, and then to further examine the heart.

Hypertrophy of the heart is not uncommon in race-horses, from long-continued exertion ; and in young thorough-bred foals, heart disease is not uncommonly seen as the result of hereditary influence. Sometimes, indeed, the disease commences in the young animal prior to its birth.

Very frequently chronic affections of the heart are left after rheumatism. Unfortunately, very little can be done in chronic affection of the valves of the heart, and indeed, in many cases, it is not our object to prolong life in horses incapable of working owing to such serious diseases.

In aged horses, after sharp exertion, rupture of the heart is sometimes met with, or it may result from a fall, or from direct violence to the chest It is, necessarily, always fatal.

CHAPTER III.

DISEASES OF THE DIGESTIVE ORGANS AND LIVER.

General remarks on the Anatomy and Physiology of the digestive organs of the horse. Acute and Chronic Indigestion, or Stomach-Staggers. Rupture of the Stomach. Colic. Lampas. Inflammation of the Mouth or Stomatitis, contagious and non-contagious. Inflammation of the Tongue. Crib-biting and Wind-Sucking. Inflammation of the Bowels. Constipation and Obstruction of the Bowels. Diarrhœa. Rupture of the Intestines. Dysentery. Diseases of the Liver. Congestion and Inflammation of the Liver.

GENERAL REMARKS ON THE DIGESTIVE ORGANS.

ALTHOUGH science has been revolutionised in our day, and the most wonderful discoveries, before not even dreamt of, have been made in physiology, chemistry, physics, agriculture, and medicine ; and although veterinary science, no less than other branches of knowledge, has made very rapid progress during the past twenty years ; yet, nevertheless, diseases caused by the grossest errors in feeding are as common as ever. We have already spoken of the diseases caused by the wilful mistakes which are being constantly made in feeding man's most faithful servant, the noble creature to which he owes a heavy debt of obligation. We have now to speak in particular of the diseases of the stomach, and these are almost invariably due to avoidable sources of disorder. We hear a great deal now-a-days regarding the ills man inflicts on himself by his neglect of the ordinary laws of physiology, dietetics, and hygiene. It is very essential also to attend to the laws of dieting of horses. They are not numerous, and are soon learnt, being, as a rule, pretty well understood by grooms who have learnt their business well. Our men should be up in the morning at 5-30 or 6 a.m., to feed the horses under their charge, and give them time to digest their food. We scarcely need mention, seeing that nearly one half of all the maladies of the horse are caused by dietetic errors, how necessary it is to bestow the greatest care and attention upon this most important subject.

There are at least four special reasons why the diseases of the digestive organs of the horse require a very careful and complete description at our hands. First, these disorders are the most commonly encountered of all equine maladies ; secondly, owing their origin in a very large number of instances to dietetic errors of one kind or another, they are the most easily guarded against ; thirdly, they are in many cases when recognised in the early stages, very amenable to judicious care and treatment ; and lastly, they are, generally speaking, very imperfectly understood.

We have already treated of several general dietetic disorders, such as weed, diabetes, and others, and now we turn to the consideration of the special disorders. Of the diet of the horse we shall treat shortly ; for it is of great importance that every owner of horses should give his attention to this important subject. Were the dietetics of the horse more generally understood, disease would be markedly diminished, more especially in the cart-horse stables.

In the horse, the intestinal tract is more liable to disease than the stomach, whereas in the ox and sheep the latter organ is more frequently affected. This is in all probability due to the fact that in the horse the stomach is much less complex than in the ruminating animals, and is also smaller in proportion to the rest of the intestines, than in the latter class of creatures. In consequence of this, the process of digestion, begun in the stomach of the horse, is largely completed by the intestines.

The digestive mechanism of the horse, and the higher animals, and man, consists of a long tube which runs through the body, beginning at the mouth, and ending at the anus. In the mouth, the food is acted upon by the salivary secretion, and is passed on into a cavity called the pharynx, which leads into the gullet. This tube passes down the neck behind the windpipe, and thence through the chest into the abdomen, where it opens into the stomach. Our readers will see at a glance how small this organ is in the horse as compared with the extensive intestinal tract. We should here mention that the body-cavity of the horse, as of all other higher animals, is divided into two halves by a sheet of muscle called the diaphragm, which stretches across from side to side. The front cavity is the chest, the hinder one is the abdomen. When the food enters the stomach, it is acted upon by the gastric juice, and it then passes on into the intestines, where it is again changed, and rendered assimilable by the secretions of the liver and those of the pancreas, and intestinal walls. When the food has passed through the various necessary changes prior to its absorption, the residue passes onwards, and is expelled at intervals from the system.

The horse is a herbivorous animal, and owing to the large amount of food which has to be taken by it, as by other creatures feeding upon vegetable matter, in order to obtain the necessary amount of nutrition, the digestive tract must present a large area for absorption. Dogs, cats, and other animals which live upon flesh do not need to eat so large a bulk of food in order to obtain the necessary sustenance ; and hence, consequently, their digestive organs are less liable to suffer under domestication, than are those of the herbivorous animals.

The little tadpole feeds upon vegetable matter, and he has a very long intestinal tract ; but when he becomes metamorphosed into a frog, which is a carnivorous creature, the digestive canal is transformed also, and the intestines become much shorter.

Professor Williams points out the important fact that easily-digested food taken by animals in excess is liable to derange the smaller intestines, whereas coarser and more indigestible food containing much woody fibre, as over-ripe hay, rye grass, and coarse straw, is more apt to accumulate in the large intestines, causing disordered action, inflammation, or even paralysis of the intestinal muscular tissues. Boiled food also is apt to be retained in the stomach, and if given in excess may cause distension, inflammation, paralysis, and even rupture. It is not only the bad quality of the food, which may set up disorders in the alimentary tract ; but irregularity in diet, and full feeding after exhausting work, are also very liable to induce disease.

The average capacity of the stomach in a horse of ordinary size, is from three to three and a half gallons; but it varies greatly according to the bulk of the animal, its breed, and the nature of its food. Relatively, it is more considerable in more coarsely bred horses, and in the ass and mule. When empty, its average weight is between three and four pounds (Chauveau). The accompanying picture shows the general shape of the stomach of the horse. The left hand opening is that of the gullet. That of the intestine is on the reader's right.

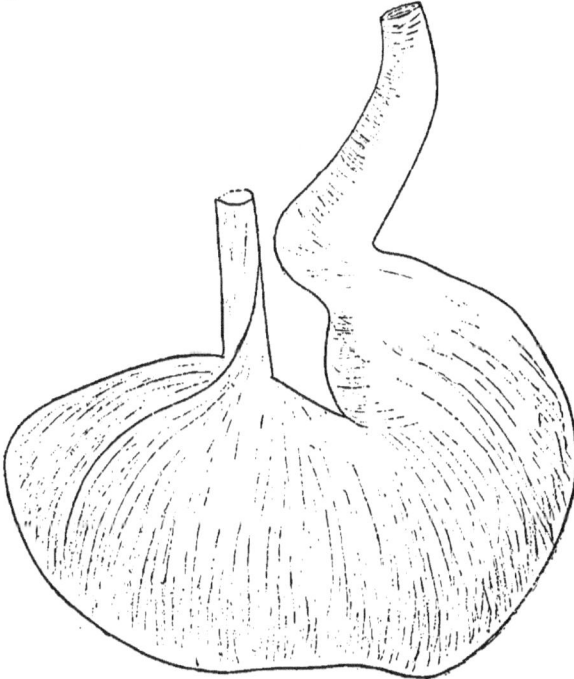

Outside, the stomach is seen to be of uniform appearance; but if the interior is opened, one is at first struck by the different aspect, which its lining membrane presents, according as it is examined on the right, or on the left side. To the left it has all the characters of the lining of the gullet, in being white, harsh, and even resisting ; and it is covered by a thick layer of cells, called epithelial. To the right it is thick, wrinkled, spongy, very vascular, and has a reddish-brown tint, which is speckled by darker patches. Here it loses its harsh consistency, and is deprived of the remarkably thick epithelial covering which it exhibits on the left side, being covered by a very thin layer of epithelial cellular structure. It is not by an insensible, but by a sudden transition that the lining membrane of the stomach is thus divided into two portions. The right part constitutes the true stomach of the horse, as on it alone devolves the secreting function, and the elaboration of the "gastric juice," the fluid which digests the food. The left part is considered to be a dilatation of the gullet (Chauveau). The accompanying picture shows this arrangement of the interior of the horse's stomach, from Chauveau's anatomy (for the English translation of which we are indebted to Dr. Fleming, LL.D., F.R.C.V.S.)

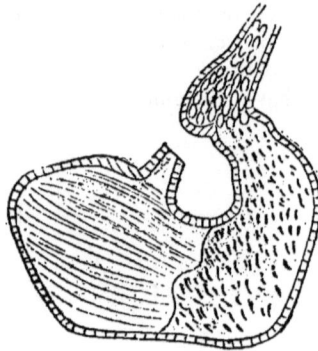

In the wild condition, the horse lived on the grass of the field ; and the smallness of the stomach, is in itself sufficient evidence that the organ is so constituted as to require to be frequently replenished in order to duly nourish the animal.

We have heard a great deal of Mr. Darwin's theory, and the theory of evolution generally, of late years, and we learn from it that "the changing conditions of the environment must produce corresponding changes in the *structures and functions* of organisms; and there is supplied in the variability of species a safety valve by which organisms which can most completely adapt themselves to the changed conditions are far more likely to survive and prosper than others less capable." (G. Gresswell, on the Evolution hypothesis.) Under domestication the conditions of life of the horse are necessarily changed. The main bulk of his food is given to him dry, and the functions of the stomach must thus be, to some extent, altered in accordance with this and other unavoidable changes.

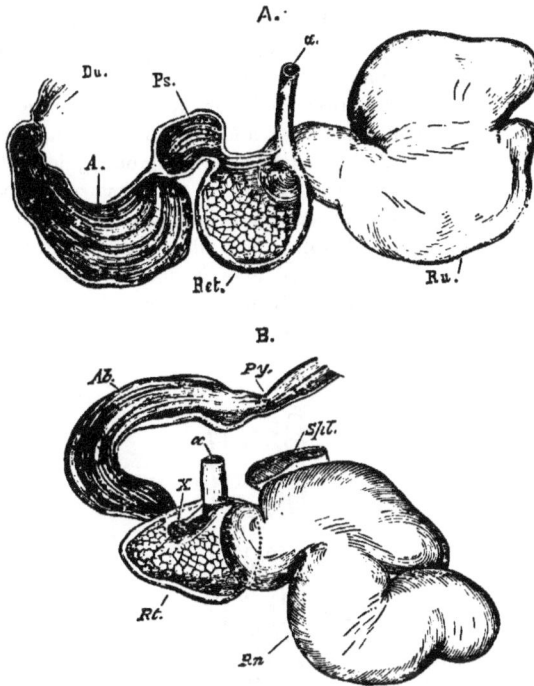

A, THE STOMACH OF A SHEEP; B, THAT OF A MUSK-DEER (*Tragulus.*) *œ.*, œsophagus ; *Rn.*, rumen ; *Rct.*, reticulum ; *Ps.*, psalterium ; *A.*, *Ab.*, abomasum ; *Dn.*, duodenum ; *Py.*, pylorus. (*After Huxley.*)

When out at grass, the horse has plenty of time for feeding; and likewise in the stable, he requires ample leisure for this purpose. We must remember that, the drier the food, the more saliva is secreted, and the longer it is retained in the mouth, in order that the starchy material contained in it may become converted into soluble sugar. The equine tribe, unlike the bovine, cannot ruminate. The ox, having filled the mouth, bestows little care upon the comminution of the food, by which means the herbage is formed into a pellet. The jaw is moved twice or thrice, and the mouthful is forwarded at once to the rumen. This receptacle is large, and is somewhat hastily filled. Then the ox retires to a quiet spot, and there enjoys the meal, the grass being regurgitated and fully masticated, during which time the animal is said to be chewing the cud. The horse has no such power. The food eaten must be well masticated, before it enters the stomach, and for this, time is required (Mayhew). It is well known that the Tartary horses are trained to undergo prolonged fasts, and to live on small quantities of food, and they are, in consequence, among the hardiest animals of the whole equine tribe. They have been specially trained to their mode and habits of life, and artificial selection by man of the hardiest dams, and the dying off of the weaker animals, have contributed to this result. Horses should, however, not be expected to undergo prolonged fasts in our country. Moreover, it is to be borne in mind that in Tartary the animals are not given large quantities of oats, hay, and other fodder in abundance immediately after exertion. They are carefully managed, and are hardy in consequence.

When food is swallowed, the gastric glands pour out a juice—the "gastric juice"—and this, unlike the saliva, which acts on the starchy foods, acts on the albuminous constituents, and renders them capable of absorption through the intestinal walls. It also dissolves the albuminous coatings of the fat cells, and liberates the fat to be acted upon by the bile. The little picture shows the structure of these glands which secrete the juice in man, magnified about 350 times. The large round cells are those which secrete the juice.

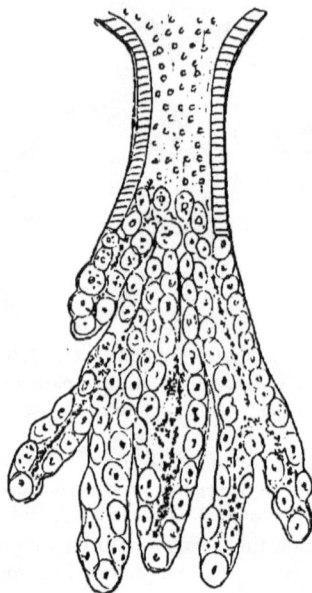

When we consider the changes the food has to undergo in the mouth and stomach before it can be absorbed, it will be seen that two hours should be allowed before a horse is worked after being fed. There is one more fact which we must mention, before passing on to the disorders of the stomach, and it is one which clearly indicates that the structural peculiarities of the horse show us that he is by nature a constant feeder. In man ; in the ruminating animals, such as sheep, camels, deer, oxen, giraffes ; and in the carnivora, such as the dog, cat, lion, tiger ; and many other animals, the bile, which is a fluid secreted by the liver, is collected in a little bag or sac, called the gall-bladder, and is afterwards poured out into the small intestine to act on the food, as the latter is being carried onwards, past the opening of the gall-duct, into the small intestine. Now in the horse there is no gall-bladder, and so the bile flows constantly and directly into the first part of the small gut. When the horse is worked immediately after a meal, the blood from which the digestive juices are directly or indirectly drawn, is required to repair the loss of tissue caused by the waste of muscular elements. Every time a muscle acts, work is done, and there is a waste of tissue, and this waste has to be repaired. Therefore, if an animal be worked immediately after a meal, the food remains undigested, causing irritation of the stomach and intestines, and various diseases, such as colic, stomach-staggers, and other affections result.

ACUTE INDIGESTION, OR STOMACH-STAGGERS. CHRONIC INDIGESTION. GASTRITIS.

THE first disorders of the stomach to which we shall draw attention, are acute and chronic indigestion. Acute dyspepsia, or indigestion with engorgement, popularly termed stomach-staggers, although not uncommon in some parts of the country, is rather rarely met with in North Lincolnshire. It results from engorgement of the stomach with food, from imperfect mastication, and from eating indigestible material, or food specially apt to undergo fermentative changes. Cooked food, brewers' grains, musty hay, and ripe vetches, are especially liable to cause impaction. Wheat and barley are also very likely to induce indigestion, and they frequently also cause purgation and laminitis, and may even lead to a fatal result. Horses are more liable to dyspepsia after severe or prolonged exertion, especially if the food be difficult of digestion, or in too large quantity. The symptoms of acute indigestion are generally sudden in their onset. There is fulness of the abdomen, and the horse is restless, and shows indications of colicky pain. He lies down and rises again alternately, and paws the ground with his fore feet. Eructations of wind, occasional discharges of saliva from the mouth, and tremblings, especially in the muscles of the left shoulder, are also among the symptoms of acute indigestion. Not uncommonly, actual vomiting occurs. Vomition is thought by the general public not to be possible in the horse. This is a great mistake. It is by no means very uncommon. Recently we had under treatment a case of acute indigestion, caused by the rapid eating of a very large amount of fresh clover, and the animal vomited a large quantity of green liquid, which passed through the nostrils and mouth. In severe cases, the pain is very acute, and the horse throws himself about wildly, and frequently looks towards his flanks. The pulse and respirations are accelerated, and in some instances the horse, instead of manifesting pain, remains dull and semi-comatose, and the breathing may become stertorous. He refuses his food, is moved with difficulty, and attempts to press his forehead against a wall or tree, or anything which comes in his way. Under these circumstances, the respiration is much quickened, and the pulse is of full volume, but not so accelerated as when gastric pain is a prominent symptom. Sometimes the disease is directly traceable to eating bad oats, especially mouldy ones.

In mild cases of stomach-staggers, the animal always recovers, and many recover without any treatment. In severe cases, the prognosis is not so favourable, as death sometimes ensues from rupture of the stomach, owing to the great distension of this organ, or from inflammation. Sometimes, but rarely, after death, calculi of oat hairs are found in the stomach.

In the treatment of acute indigestion, a purgative should be given at once. For this purpose five or six drachms of good aloes is preferable to any other aperient, though oil is recommended by some. If there be much flatulence, we may give an ounce to an ounce and a half of aromatic spirit of ammonia in half a pint of gruel. If there be any pain, an ounce to an ounce and a half of sulphuric ether, with half an ounce of spirit of chloroform, may be administered in water or gruel every two or three hours, as long as the pain continues to be severe. In ordinary cases the spirit of ammonia, spirit of chloroform, and sulphuric ether answer well together; and when given three or four times daily, will be found very beneficial. When gastric irritation is great, we may add thirty minims of diluted hydrocyanic acid to each draught. If the aloes does not act within the first forty-eight hours, it is best to administer one to two pints of castor oil. In stomach-staggers, it is never advisable to abstract blood. In the further treatment of this disease, as soon as the appetite returns, the diet should be at first of a laxative nature, and limited in amount.

Chronic indigestion, like the acute form, is chiefly caused by dietetic errors, though these are not always apparent. Sometimes the food, though of good quality, is too stimulating and dry. In other cases it is not sufficiently masticated, perhaps owing to irregularities in the teeth, but in most cases it will be found to be of inferior quality, or administered irregularly.

Dietetic errors induce changes in the gastric juice, and in the movements of the stomach, which, however, are sometimes deficient from impaired nerve power. The symptoms of chronic indigestion are very variable. The appetite may or may not be impaired. Sometimes it is capricious and perverted. In other cases, though it continues good, the animal still continues to lose flesh. The bowels are generally irregular, the fæces often coated with mucus, and there may be great thirst and acid eructations. Abdominal pain is not unfrequent in severe cases, especially when the appetite remains unimpaired. The horse is weak, sweats easily, and the skin is dry and hard. These cases are but too frequently dosed and poisoned with over-doses of aconite drenches. In treating chronic indigestion, the causes of the disorder should be enquired into, and the dietetic arrangements carefully regulated.

Some purgative should be given in all cases, unless the bowels are freely open, and should be followed by the administration of vegetable tonics. The diet should be limited in amount, and it will be advantageous to change it. Internally we may administer half an ounce of Fowler's solution, with half an ounce of bicarbonate of potassium in the drinking water, twice or thrice daily after meals, for three or four days. Afterwards we may administer stimulating balls, eight drachms each, of equal parts of carbonate of ammonium, ginger, and gentian, made up with treacle. These may be given at first twice, then once daily. Inflammation of the stomach or gastritis may be acute or chronic. It is commonly due to toxic agents taken, but may come on from indigestion, or it may arise from foreign bodies such as calculi, or be due to bots. Crib biting also is not unfrequently a cause of dyspepsia and chronic gastritis. The treatment of these affections is in the main similar to that of the preceding. Under any circumstances the food should be restricted in amount in the acute variety, consisting of linseed gruel and other non-irritating material ; but in all cases it is necessary to ascertain the cause, and if any poison has been ingested, it will be necessary to treat the inflammation in accordance with the nature of the toxic agent. Tincture of opium in doses of one ounce and a half, with one ounce of sulphuric ether, and one ounce of spirit of chloroform, given three or four times during the day in three quarters of a pint of gruel, will be found very beneficial Fortunately, acute gastritis is not very common in the horse.

RUPTURE OF THE STOMACH.

THERE remains for our consideration but one more disease of the stomach, and although it is nearly always fatal, we propose to treat of it pretty fully, because it illustrates so well the baneful effects liable to be caused by injudicious feeding and work. Partial or complete rupture of the walls of the stomach is not uncommon among horses. It is mainly due to errors in dieting and work, and is more frequently met with among the heavier draught horses, which are especially subjected to irregularities of work, and defective dietetic arrangements. Rupture is especially likely to occur when a large amount of food is given after exhausting or prolonged work. Under these circumstances, the food is especially liable to undergo fermentative changes from its longer retention in the stomach, owing to the slow and imperfect action of the gastric juice and defective movements of the walls. It is more frequent in horses fed on bruised than on whole grains, especially when put to work after a full meal. Rupture of the stomach is probably, in most instances, preceded by derangement and distention, or actual disease of the walls, consequent on chronic indigestion and other causes. It is rarely met with in young animals, but most commonly occurs in aged horses, especially when these have undergone severe exertion, or have been overworked for a long period.

Vomiting, or attempts at vomiting, generally occur in rupture of the stomach, but as it is not invariably present, and may proceed from other causes as, for instance, rupture of the large intestine, it cannot be considered a distinguishing symptom. Vomition is, however, more complete in rupture of the stomach, and in dilatation of the opening of the gullet into it, than in rupture of the intestine. Rupture may be brought about by gradual distension of the walls of the stomach, and without much pain, until the contents escape into the body-cavity. Sometimes, however, rupture occurs suddenly, owing to the violent struggles of the animal in its paroxysms of pain, during the course of disease of the stomach or gut. The animal becomes uneasy, with countenance dejected, and he looks anxiously round at his flanks. There is great weakness and rapid prostration of strength. The pulse is feeble and fluttering, the respirations are short and quick, and there are frequent attempts at vomiting. In some cases the animal remains quiet for a time after the rupture, while in other instances the pain is intense, and the animal becomes delirious. In some cases of rupture, collapse and death follow in a few hours; while in others, where the rent is not so extensive, life may be prolonged for a couple of days or more. Treatment is of no avail, and if the veterinarian decides the case to be one of rupture, he deems it best to have the animal put out of his agony.

Our readers will perceive that cases of chronic indigestion, the treatment of which we described above, may, if neglected, lead eventually to rupture of the coats of the stomach. We may conclude our observations on the diseases of the stomach by remarking that in the horse a staring coat, sluggishness at work, emaciation, with a tucked-up appearance of the belly, are among the most apparent signs of dyspepsia. The presence of undigested food in the fæces, and especially of un-crushed oats, and the occasional appearance of griping pains, all indicate that the digestive organs are at fault (Gamgee).

COLIC.

PAIN in the abdomen may arise from derangement of the functions of the intestinal tract, or it may be due to actual organic changes of varying extent and nature. To the former disorder the term "true colic" is applied, while, when depending on organic disease, this condition is sometimes spoken of as "false colic." "True colic" is of two varieties, which may be associated together. The one termed "spasmodic colic," is due to spasmodic contraction of the muscular wall of the gut ; the other, termed "flatulent colic," is owing to extensive gaseous accumulation in the intestine. It is said that the spasmodic form may terminate in intestinal inflammation. Colic is generally due to dietetic errors, such as overfeeding, irregularities in the diet, such as food of inferior quality or unsuitable kind, taking a large amount of food after a long fast, or it may arise from prolonged or severe exertion. These causes, are all potent agencies in the production of spasmodic intestinal contraction, as well as of flatulent distension, which may either be associated with the spasmodic variety or occur independently.

Besides dietetic errors, there are many other causes of colic. This painful affection may be due to mechanical displacement of the bowels, or to the presence of parasites in the bowels, and sometimes also in the neighbouring parts. Young animals, especially when badly fed and attended to, are more liable to colic from this source than older horses.

The onset of spasmodic colic is generally more or less sudden. The horse shows signs of abdominal pain by looking round at his flanks, by restlessness, by striking at his belly with his hind feet, and in various other ways. He lies down and rolls about from side to side. After a while he rises and eats a little, and soon, perhaps, a paroxysm of pain again attacks him.

In uncomplicated cases of colic, the pulse and respirations and temperature are rarely elevated, except during the paroxysms of pain. The pulse is then much accelerated, and the respiration usually becomes hurried. The attack may now subside, or may gradually become more and more severe, the paroxysms being more continuous, and the pain more intense. The restlessness and excitability increase, and partial stupor supervenes. The attack, if unrelieved, may end in death from continued pain or from exhaustion, with varying complications. In most cases of colic the bowels are constipated, and the excrement, if any, which is passed is usually hard and often coated.

The flatulent form of colic due to distension of the intestine with gas, may be associated with spasm of the muscular coats, or it may occur independently of it. It is especially to be attributed to digestive disturbance depending on ingestion of food, which is prone to undergo fermentation. This affection usually comes on suddenly. The horse is noticed to be very restless, and the abdomen distends and becomes tense, and gives a hollow note if struck. The breathing is short, and the pulse is frequent and feeble. The extremities are cold, and there may be more or less delirium and dizziness. When the animal lies down, he does not throw himself suddenly on the ground, as in spasmodic colic, but allows himself to fall more slowly and carefully. If unrelieved, the continued distension may lead to further disturbance of the heart and lungs, and death may result from asphyxia. Sometimes rupture of the colon or other part of the bowel or of the diaphragm is the cause of death.

In ordinary cases of colic, the prognosis is very favourable in both varieties, but it is not so good in severe cases of the flatulent kind. In all prolonged cases with great pain and restlessness, there is danger of displacement or entanglement of the intestine; and when gaseous distension is very great, and the struggles are very violent, there is great risk of rupture of the large bowels or of the diaphragm.

In inflammation of the bowels, of which we shall treat hereafter, the abdominal pain, unlike that of colic, is continuous, it is more agonising, and rarely has periods of intermission. The prognosis in this latter disease is very grave.

In all cases of colic, except those in which diarrhœa is present, it is

F

advisable to commence treatment by administering a purgative medicine. Barbadoes aloes is the best purgative, and is given preferably in the form of a ball, in doses of five, six, or seven drachms, depending upon the size and condition of the animal. It is of the greatest importance that the aloes be of the best quality. Much of that sold is really unfit for use, and great care is therefore necessary in procuring the drug. In all cases where a full dose of aloes is given, three full days' rest is absolutely necessary. In addition to the cathartic, clysters of water at about 100° F. should be given, and repeated at intervals of two or three hours if necessary. In addition, a drench composed of one ounce of sulphuric ether, one ounce of tincture of opium, one ounce of nitric ether, and half an ounce of powdered pimento, should be given at intervals of one or two hours, in a pint of gruel, as may be necessary. Some recommend Fleming's tincture of aconite, but this remedy is not necessary in the treatment of simple colic.

Hot fomentations, and stimulating liniments, or mustard, may with advantage be applied to the abdomen. The diet should consist of warm water and bran mashes.

In flatulent colic it is advisable that the drenches should also contain one ounce of aromatic spirit of ammonia.

LAMPAS.

WE will now turn to the consideration of the diseases of the mouth, and in this connection shall treat of lampas, inflammation of the mouth, inflammation of the tongue, crib-biting, and wind-sucking.

Some of our readers will be a little surprised to hear that lampas is not a disease at all. It is merely a swollen condition of the palate, occasioned by a determination of blood to this part, which is the seat of active changes, during the development of the teeth in young horses. Gamgee, in his work on the domestic animals in health and disease, states that whence the absurd name lampas is derived, he cannot venture to determine ; but he observes that it has done much mischief, by being regarded as a specific name for a specific disease, *supposed* to require active treatment by the *hot iron*. In many instances this swollen condition of the palate which manifests itself by projecting below the level of the front teeth requires no treatment. Sometimes lampas occurs as a manifestation of disorder of the stomach or intestinal organs. When it is due to a congested condition of the gum, occasioned by teething, a few pricks with a lancet, or bathing the part with an astringent solution, consisting of two drachms of alum and one ounce of tincture of myrrh to twelve ounces of water, will relieve the irritation. The latter method of procedure is much the best for the amateur, as it is nearly always sufficient. Care must be taken in lancing the gum not to cut the artery underneath. The animal should be fed for a few days on mashes, and half an ounce of bicarbonate of potassium with two drachms of nitre, may be given in the drinking water twice daily.

When lampas is due to digestive derangement, this condition must be treated. Locally, nothing need be done, but the same internal treatment may be adopted as when teething is the cause. It is well in all forms to commence treatment by the administration of a mild dose of aloes, say two to four drachms.

INFLAMMATION OF THE MOUTH OR STOMATITIS, BOTH CONTAGIOUS AND NON-CONTAGIOUS.

INFLAMMATION of the mouth or stomatitis, occurs in several forms in the horse ; and in addition to the more common varieties, there is a contagious disease of the mouth in which little blebs, which eventually become pustules, are formed on the tongue and on the membrane lining the inside of the mouth. This disease, however, is so rarely met with as to require only a cursory notice. Recently we have had two cases under treatment. In these cases the whole of the lining of the mouth underneath the tongue and on the inside of the cheeks was found to have peeled off in flakes, and the raw surface left was studded copiously with little rounded ulcers. The mouth was hot, dry, and red. This disease is specially interesting from the fact that it is so very readily communicated to man. In these cases of which we are speaking, both the foreman and the shepherd became inoculated while giving balls prior to calling in the writer, and both became seriously ill, but they eventually recovered. We will now consider the simple non-contagious varieties of inflammation of the mouth. This disease is most frequently met with in young animals, which are especially predisposed to it by malhygienic conditions and improper dieting. It may be due to local irritation or to mechanical causes, or may follow disorders of the digestive organs. It often occurs in animals debilitated by disease, and is in many instances traceable to a certain vegetable fungus, or to a special acarus.

There are several varieties of non-contagious stomatitis. The first called simple stomatitis is chiefly met with in foals. The first manifestations of this disease are small circumscribed red patches on the membrane lining the cheeks and roof of the mouth. These patches are covered with a yellowish film, which soon separates, and leaves a superficial erosion. The breath smells badly, and thick saliva accumulates in the mouth. This form is sometimes seen in old animals, and is often associated with bad digestion. The next form is called vesicular. In this affection small blebs appear on the membrane lining the mouth, and some of these rupture. In the third or last form, called pustular, the vesicles become pustular. The hygienic and dietetic arrangements should be attended to in the first place in stomatitis. As local applications, lotions of chlorate of potassium, twenty-five grains to the ounce of water, or of carbolic acid one part in forty of water are very useful. If the ulcers become very offensive or indolent, they may be painted with a solution of nitrate of silver, ten grains to the ounce of water, once daily. A mild purgative should be given in the first instance, and half an ounce of bicarbonate of potassium, with half a drachm of chlorate of potassium may be given in the drinking water twice daily, so long as the disease lasts.

INFLAMMATION OF THE TONGUE.

INFLAMMATION of the tongue is not often seen in the horse, except when resulting from the action of irritants or from mechanical injury. It may follow on inflammation of the structures near it. The tongue becomes swollen, hot, tense, and painful, and soon protrudes in consequence of its increase in size. There is difficulty in swallowing, and saliva and mucus accumulate in the mouth. Gargling firstly with warm water, and afterwards with carbolic acid lotion (one in forty of water), or alum lotion (four drachms in eight ounces of water), is very useful. We spoke, in treating of anthrax, of a disease called glossanthrax, and this rare malady, we said, was nearly always fatal. It has nothing to do with simple inflammation of the tongue, though one of its chief manifestations is the enormous size which the organ assumes.

CRIB-BITING AND WIND-SUCKING.

CRIB-BITING is a habit in which the horse seizes hold of the manger with his teeth, and forces out wind from the stomach. " In wind-sucking the horse smacks his lips, gathers air into his mouth, extends his head or presses it against some solid body, arches his neck, gathers his feet together, and undoubtedly swallows air, blowing himself out sometimes to a tremendous extent " (Williams). As may be imagined, the latter vice is more serious in its consequences than the former, though both constitute unsoundness.* They both cause digestive derangement, but it seems not improbable that wind-sucking may be an effect as well as a cause of indigestion and impaired general health. Enforced idleness is one of the causes of these habits. Other factors are dyspepsia, and imitation of animals addicted to these vices. Crib-biting wears away the foremost edges of the central and lateral incisor teeth, and by this worn appearance of the teeth, the habit of the horse is easily detected. There are various ways in which these vices may be prevented. The most common method is to place a strap round the neck loose enough to allow the swallowing of food and drink, but too tight to admit of the muscles of the neck being tightly contracted.

---

*As we shall frequently have occasion to refer to the question of unsoundness, it seems advisable to give our readers the legal interpretation of the term. It is as follows :—
" If at the time of sale the horse has any *disease* which either actually *does diminish* the natural usefulness of the animal, so as to make him less capable of work of any description, or which in its ordinary progress *will diminish* the natural usefulness of the animal, this is unsoundness ; or if the horse has, either from *disease* or *accident*, undergone any *alteration* of *structure*, that either actually *does* at the time, or in its ordinary effects *will* diminish the natural usefulness of the horse, such a horse is *unsound.*" (Lord Ellenborough in the case of Elton *v.* Brogden, 4 Camp. 281).
It will be seen from this, that the term *unsoundness* is an extremely elastic one, and therefore that the very greatest care should be taken in pronouncing a horse *sound.*

Wind-suckers and crib-biters should have a supply of water by them, as well as a good lump of rock salt in the manger. Half an ounce of bicarbonate of potassium given-in the water once daily, will sometimes prove very beneficial.

Inflammation of the gullet is generally the result of direct injury. It may be induced by the passage of very large portions of food, or by the administration of irritant liquids. The symptoms are difficulty of swallowing, tenderness on manipulation, and sometimes spasm of the gullet. Many cases do not require treatment beyond the use of liquid food for a day or two. In severe cases, linseed gruel, to which an ounce of tincture of opium has been added, is of benefit where there is much pain. Hot water fomentations are useful.

INFLAMMATION OF THE BOWELS.

WE now turn to the consideration of inflammation of the bowels, but may first say a few words about inflammation generally. What is inflammation? Inflammation comprises three kinds of changes. Firstly, there is a change in the blood vessels, and in the circulation through them ; secondly, there is an exudation of fluid, and often little blood cells escape through the walls of the vessels ; and lastly, there is change in the tissues themselves. The arteries first enlarge in inflammation, and the blood flow is accelerated. Afterwards, the blood flow is much retarded, and little cells of the blood accumulate in the small veins, and stick fast in those little communicating tubes between the veins and arteries, which are called capillaries. The blood flow through these channels at last stops, and then some of the little blood cells pass out of the containing vessels into the parts around. Inflammation leads to depression of vitality, and death of the tissues involved. There is no increase of vitality, and no multiplication of the elements of tissue.

Inflammation of the bowels or enteritis in horses is a disease of very great importance, being both very fatal and very common. It is more commonly met with in adults, and in those animals which are in confinement, than in the young and those out at grass. It is also of more frequent occurrence among the heavy draught horses, than among the more highly bred animals. This fatal malady has two distinct forms, which, although presenting many symptoms in common, are in reality of a different nature.

The first variety we may term apoplectic, from the rapidity with which the animal is struck down. The horse may even die in a few hours. Of this variety the causes are not always apparent. Over-exertion, prolonged exposure to cold, drinking cold water when heated, and, finally, washing the animal in cold water, while still in a heated and perspiring condition, are however the most important. This variety is generally fatal. The other variety, which may be termed secondary inflammation of the bowels, is in most instances of not such a severe type, even though the extent of inflammation is sometimes very great. This form is due to continued obstruction of the bowels from various causes, or to taking irritant poisons such as arsenic, or to irregularities of feeding and work.

In inflammation of the bowels, sometimes the large and sometimes the small intestines are invaded, while at other times both are involved simultaneously. In some cases the symptoms of enteritis are gradual, while in others they are sudden in their onset. Not unfrequently the pain in the belly is preceded by general constitutional disturbance, shown by acceleration of the breathing, marked dulness, depression and loss of appetite; while, in other cases, the inflammatory action is ushered in with marked shivering or rigors.

Inflammation of the bowels is a disease very prevalent among the heavy cart horses, not only in the country districts but also in our large towns; and it is therefore very important that its general symptoms should receive the special attention of horsemen. Our reason for laying stress on this matter is, that the disease in question is one of all others which is especially amenable to *early*, judicious, careful treatment. The belly in enteritis is very tender when pressure is applied, and in this particular we have a feature which helps us in diagnosing the disease to be one of far greater danger than simple colic. The pain also, unlike that of simple colic, is continuous, is more agonising, and but rarely has periods of intermission. The pulse, at first quick, hard, and wiry, becomes in the later stages still more accelerated, though of less volume, feebler, and gradually more irritable and imperceptible. In number, the beats range from 70 or 80 to 120 per minute. In cases of simple colic, the pulse is unaltered, except during the paroxysms of pain; whereas in enteritis it gradually becomes more and more disturbed. The animal in the paroxysms of pain, stamps and strikes at his belly, and when he lies down, he may be observed to do so with greater care than in simple colic. He often turns his eyes towards his flanks. Copious sweats bedew the body. He groans in agony. At other times he stands almost motionless, with an expression indicative of acute suffering depicted on his countenance. The surface of the body becomes cold, the pupillary openings of the eyes dilate, and delirium and stupor may supervene. The animal soon, perhaps, becomes more restless than ever, and wanders about the box, or casts himself down, and rolls about regardless of obstacles. Sometimes the animal will balance himself for a short time, with teeth clenched, and limbs and ears very cold, when he may suddenly fall, and die exhausted after severe struggles. Some time before death, an apparent improvement may take place. The horse stands quiet for a while, yet, though the breathing becomes quieter and the pain abates, and he takes a little food, the countenance maintains its haggard, dejected appearance, cold sweats bedew the body, and the pulse continues to be thready, and perhaps almost or quite imperceptible. In still more advanced stages, if agony, pain, and intense inflammation have not already carried off our patient, he trembles continuously, the lips fall pendulously, the eyes become

duller, the mouth becomes clammy, the breath perhaps fetid, until at length
he can hold out no longer, and death puts an end to his suffering.
The mortality in enteritis varies from 45 to 65 per cent. If, as happens
in some rare instances, the acute symptoms abate after the lapse of a few
hours, and the pulse regains in some degree its normal character, becoming
fuller, softer, and slower, there is great hope of recovery.

In the form of enteritis, which we spoke of as apoplectic, the appearances
found at the autopsy are very marked and characteristic. The lining
membrane of the affected section of the gut is intensely congested, being of
a deep purple or even black colour, and in many instances much blood is
effused into the intestinal canal. The lining membrane is also much
thickened, and can easily be separated from its connections with the
underlying coats of the gut. The other coats are also intensely infiltrated
with blood-stained effusion. In some cases so extensive is the infiltration
and thickening, and so intense is the inflammatory process, that the tissue
just outside the lining membrane appears as a dark purple or black gelatinous
mass two inches or more in thickness, extending for varying lengths of the
gut, and sometimes involving many feet of the intestinal tract.

It is noteworthy that even though the amount of effusion into the gut be
very great, and the contents themselves be fluid, the bowels usually remain
inactive, owing to paralysis of the muscular coats. In other forms of
enteritis the inflammatory process is not of this marked character : the
inflammation is usually more patchy in distribution. Inflammation of the
bowels requires all the care and attention of the high-class veterinary
surgeon.

In cases of enteritis, a drench containing seven minims of Fleming's
tincture of aconite, two drachms of chloroform, one ounce of sulphuric ether,
and one ounce of tincture of opium, given in a pint of gruel or water is an
efficacious mixture. It may be repeated at first every two hours for four or
five times, and then every four hours, so long as the pain lasts. It is of
primary importance in all cases of inflammation of the bowels to control the

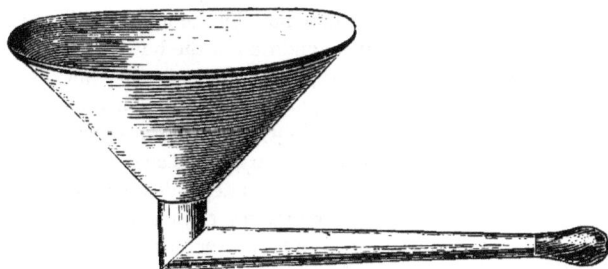

CLYSTER PIPE.

pain by the administration of such anodynes as these mentioned, for the
continual struggles of the animal often lead to rupture of the gut, which is
necessarily followed by death. Belladonna is not of much value in the

treatment of enteritis in horses, and should not therefore be administered. Hot fomentations by means of woollen rugs wrung out from very hot water, should be applied to the belly and renewed every half-hour for five or six times, while the pain is very acute, and afterwards every hour or so. During the time when the rugs are being renewed, some stimulating liniment may be well rubbed in by the attendant, over the belly. Some practitioners prefer the application of a poultice of mustard, which is rubbed off in two or three hours, and followed up by the application of hot fomentations. Enemas of tepid water should be given by means of the ordinary funnel apparatus, but on no account is it advisable to use an injecting syringe.

If the horse is inclined to drink, he may be allowed linseed gruel or linseed tea, or thin oatmeal gruel. We do not recommend the use of purgatives in enteritis. After the abatement of the acute symptoms, the diet should be laxative, consisting of bran-mashes, linseed and oatmeal gruel. No hard food should be allowed on any account until all danger is over. In some animals in high condition, bleeding is indicated in the early stages. Blood, however, should be abstracted in moderation only. It is our practice never to remove more than two or three quarts, and never to repeat the operation.

CONSTIPATION AND OBSTRUCTION OF THE BOWELS.

CONSTIPATION, or torpid action of the bowels, is by no means uncommon in the horse ; but, although it very seldom leads to a fatal result, it nevertheless deserves attention and judicious management. It depends upon obstruction of the bowels, or upon deficient intestinal action or secretion. The two latter are in their turn chiefly due to dietetic errors, though they may also depend upon other causes. Generally, the belly is full and distended, but this is by no means a constant symptom. If the constipation continues unrelieved, the appetite fails, weakness follows, and the pulse becomes feeble and accelerated. In some instances a mucous secretion is discharged in cases of constipation, and this is frequently mistaken for diarrhœa by the uninitiated, when, on the contrary, it is indicative of a costive condition of the bowels. As long as the animal remains in pretty good health, all that is necessary in constipation is a more laxative diet. If the constipation is habitual, a moderate dose of aloes, say four or five drachms, followed up by the administration of vegetable tonics, such as nux vomica, gentian and others, is efficacious in most instances.

In some cases irregularity of the bowels depends upon paralysis of part of the intestine, and in these cases purgatives cannot be administered. In these cases a mixture consisting of liquor strychninæ hydrochloratis two drachms, and of aromatic spirit of ammonia one ounce, may be given three times daily in half a pint of gruel. For the prevention of the recurrence of constipation, bran-mashes and other laxative diet may be substituted occasionally for the more solid food ; and eight drachm balls made up of equal parts of carbonate of ammonium, ginger, and gentian made up with treacle, with the addition of one grain of sulphate of strychnine, may be given three or four times weekly with great advantage. If necessary, an occasional dose of aloes may be given. ·

In cases of young foals unable to void the excretions, clysters of oil will generally be found efficacious. If the bowels are not relieved by this means, or if it is found inconvenient to adopt the above measures, two and a half ounces of castor oil may be administered internally.

There are many causes of obstruction of the bowels in horses. Sometimes it is due to impaction of matter in the gut, such as concretions of various kinds. Sometimes it is due to twist of the gut, and again at other times it is owing to passage of one part of the bowel into that immediately below it. The symptoms of obstruction which may proceed from so many different causes are very variable. When it is due to impaction of faeces or to the presence of concretions, the symptoms are usually gradual, and of an intermittent character. The other forms are generally more sudden in their onset. Twists or strangulations of the intestines of the horse are generally associated with great abdominal pain, restlessness, sitting on the hind-quarters, small, frequent, thready pulse, accelerated respiration, cold extremities, distended belly, and collapse ending in death from exhaustion. In these cases where there has been constipation with frequent attacks of colic, the obstruction is probably due to impaction of faeces.

In cases of absolute stoppage of the bowels, a drench, composed of one ounce and a half of sulphuric ether, half an ounce of chloric ether, and one ounce of tincture of opium, given in three-quarters of a pint of gruel or water, will be found very useful. Clysters of warm water may be injected in full amount into the rectum, and hot fomentations or woollen cloths wrung out from hot water applied to the belly. The food should be of a laxative kind, and only allowed in moderation.

In very severe cases of constipation, sulphate of eserine is an invaluable remedy, when injected in solution, intravenously.

On May 19th, 1886, a six year old strong draught horse was affected with colic, due to obstinate constipation. We were called in to see it on the 20th, and gave a six drachm aloes ball. At 4 p.m. on the 21st, there was no relief; the pulse was 96; the pain was almost continuous. From the foreman's calculations, there had been no passage for three days. One grain of sulphate of eserine was given by intra-venous injection into the jugular vein. In eight minutes there was extreme pain; the horse broke out in twenty minutes time into a profuse perspiration. Muscular tremors were marked features. At the twelfth minute after the injection, hard faeces together with fluid were passed. Violent straining continued for the next twenty minutes, during which time the animal voided no less than thirteen distinct discharges of alvine material, several of the latter being quite fluid. The tenesmus and grunts were intensely extreme. Muscular tremors increased up to half an hour after the injection, after which time the animal quietened, ate some mash, and the pulse went down to 60.

On leaving, further action was prevented by the administration of an opiate draught.

The horse is now well, and completely recovered.

We have also used this valuable salt in cattle. A cow had been constipated for four days, and had during that period absolutely no passage. She had received full doses of Epsom salts, linseed oil, castor oil, aloes, jalap, calomel, and other aperients. Half a grain of sulphate of eserine was injected into the jugular vein, with no other results, than pain manifested by frequent moanings, gurglings in the bowels, and restlessness. After an interval of half an hour, three-quarters of a grain were injected. In nine minutes the cow passed hard lumps, and fluid fæces. On the following day, three-quarters of a grain was again injected with good results. The animal had two or three passages afterwards during the day, and then made a gradual but complete recovery.

We have tried it in three or four other cases, and find that unless from three-quarters to one grain is given, and that by intra-venous injection, little or no action is produced. It is well to bear in mind that this drug should never be given except in very severe cases, as the action is so extremely excessive.

DIARRHŒA.

DIARRHŒA is the general term applied to abnormal fluidity, and increased amount of the alvine discharges. The proximate causes of diarrhœa are excessive secretion from the walls of the gut, combined with increased action of its muscular coats. These conditions are in their turn either due to direct irritation of the lining membrane from without, as, for instance, by food, foul water, parasites, or to indirect nervous influences. Perhaps, of all causes of diarrhœa, the most frequent in the adult animal are injudicious and irregular dieting. Sudden changes in the diet, especially from a dry to a moist or laxative one, ingestion of certain substances, copious draughts of cold water, when the animal is heated after exposure to the sun's rays or exertion, and feeding immediately after severe work or exposure to cold and damp, may be mentioned as specially liable to induce diarrhœa. In plethoric horses, doing very little work, a small amount of exercise will often bring on an attack of diarrhœa. In some cases of diarrhœa there is great prostration, the breathing becomes more rapid, and pain in the belly is not uncommon. The pulse is usually not much altered.

In the young, diarrhœa in many instances differs from that of the adult, having characteristic features of its own. The form of diarrhœa to which we refer is a specific catarrh of the bowels which, though not contagious in foals, as it is probably in the bovine tribe, is, nevertheless, a far more serious affection than ordinary diarrhœa of the adult. It owes its origin to defective sanitary arrangements, and also to changes in the quality of the milk. Such changes are traceable in some instances to the fact that the mare is worked hard during the day, and returns at night to her foal, which, after its fast during the day, is apt to take more milk than it can well digest. The symptoms of this diarrhœa of foals usually appear during the first two or three weeks of life. The alvine discharges are at first of a yellowish-white colour, and there is little or no pain. In more advanced stages there is more or less abdominal pain, which may be very severe. If the disease continues, the foal ceases to suck, and loses flesh rapidly. The prognosis is usually very favourable in ordinary cases of diarrhœa, but in the infantile variety a fatal termination is not uncommon.

We should, before treating diarrhœa, in the first place, endeavour to ascertain its cause. If it proceed from irregularities in the feeding, or in the work, these should be immediately rectified. In most instances medicine is not required, unless the diarrhœa is excessive, or the pain and general disturbance very great. No cold water should be allowed on any account, and the animal should be kept quiet, and warmly clad. The food should be easily digestible, and linseed gruel or other demulcent drinks may be allowed. Where there is very great pain, a drench composed of one ounce of tincture of opium, one ounce of spirit of chloroform with a drachm of camphor, may be given three times daily in flour gruel. When prostration is very marked, and the pain severe, one ounce of tincture of opium, one ounce of sulphuric ether, and one ounce of spirit of chloroform may be given three times daily, with a moderate amount of alcoholic stimulant, such as brandy or whiskey. Woollen cloths wrung out from warm water may be applied frequently to the belly, and stimulating liniments rubbed in during the intervals between the applications. In treating *diarrhœa in young foals*, it is well to commence by giving three ounces of castor oil, in order to expel the irritant matter in the intestines. With the castor oil, two drachms of sulphuric ether may be given. This treatment may be followed up by the administration of camphor and opium, with spirit of chloroform three times a day in water, or in strong decoction of tea. Each drench may contain of camphor, twenty grains ; tincture of opium, two drachms ; spirit of chloroform, three drachms ; liquid extract of bael fruit, six drachms ; and water to four ounces. When weakness is very marked, a little alcoholic stimulant may be added, and the hot cloths and stimulating liniments applied to the abdomen.

RUPTURE OF THE INTESTINES.

RUPTURE of the walls of the gut is of more frequent occurrence in the large than in the small intestines, and is due in most cases to impaction of excreta or to excessive accumulation of wind, or to both these conditions together.

It will easily be seen that these disorders are especially liable to result when the walls of the gut are in an unhealthy condition. The symptoms of rupture are very variable and not characteristic. In most cases rapid exhaustion follows the intestinal rupture, wherever it may be situated. In some cases, collapse and death soon follow the occurrence ; while in others, life is not extinguished for several days. In many cases, rupture is difficult to diagnose from several other severe affections of the bowels. Sometimes, when following impaction of the excreta in the large bowel, or great accumulation of wind, rupture is succeeded by relief. The restlessness and straining subside, and a period of calm follows, until death results. The countenance is anxious. The pulse is small and thready, and gradually becomes more and more imperceptible. The breathing is short. There is great disinclination to stir.

In rupture of the large sacculated bowel, the horse frequently sits on his haunches, and may attempt to vomit ; but these symptoms cannot be said to be characteristic. Sitting on the haunches, indeed, is a very frequent symptom in twists and other forms of disease of the bowels. Treatment is of no avail in rupture of the bowels.

DYSENTERY.

DYSENTERY is of less frequent occurrence in the horse than in the other domesticated animals, and, owing to the comparative rarity of its appearance, it is hardly necessary for us to give our readers a lengthy account of its characteristics. It has not yet been established whether this affection can be communicated from one horse to another, but it is not improbable that it may sometimes spread in this way. In most cases dysentery occurs as an independent affection, while sometimes it supervenes on an attack of ordinary diarrhœa. Among the chief causes of dysentery are overcrowding, vitiated air supply, exposure to noxious emanations, insufficient or bad food, foul water, exposure to cold and damp, overwork, and all other depressing agencies. Malarial poison arising from decaying vegetable matter, is also a common cause of dysentery. This is more especially the case in low-lying, marshy tracts, and in shady places. Sometimes dysentery begins insiduously, in which case we may at first not suspect the true nature of the affection. As the disease progresses, however, the appetite becomes more markedly impaired. Great depression and thirst, general wasting, and severe prostration are marked features. Usually the attention is first attracted by the frequency of the alvine discharges, but not unfrequently febrile manifestations, debility, and rapid prostration precede the other symptoms. These discharges are thin and watery, and are sometimes voided with great pain, and in most cases there is much straining.

Mild cases of dysentery usually terminate in recovery, but in severe ones there is not much hope of amelioration. In the treatment of dysentery, it is at first necessary to attend to the sanitary arrangements. The animal should be kept at rest, and the diet should be of an easily digestible, fairly nutritious, moist kind. A small dose of oil, say three-quarters of a pint of linseed oil, may be given in the first instance. This should be followed up by the administration of drenches composed of one ounce of tincture of opium, two drachms of camphor, half an ounce of nitric ether, half an ounce of bicarbonate of potassium, with water to ten ounces. These drenches may be given twice daily. If the progress of the disease be not arrested in a few days, astringent medicines will be necessary. In such. cases, eight drachm balls, made up of two drachms of opium, and one drachm of acetate of lead, with a sufficiency of gentian and treacle, may be given twice daily. Each ball should contain in addition, twenty drops of carbolic acid. A favourite mixture of ours is one containing sulpho-carbolate of sodium, given with tincture of opium, liquid extract of bael fruit and ipecacuanha. It is the most efficacious combination we are acquainted with.

Though we have no evidence, as yet, that dysentery owes its origin to the entry of any germ into an already unhealthy and depressed system, it seems not at all unlikely that this is the case.

DISEASES OF THE LIVER.

THE liver is the largest of all the glands. It weighs in the horse about eleven pounds. It has three very important functions. The liver of the horse is far less commonly affected by disease than that of man, in whom it is too frequently injured by immoderate drinking. We may remark, incidentally, that if the alcohol were more freely diluted than it frequently is the effect would not be nearly so deleterious, even though the actual amount taken were the same. The liver is a gland made up of oval portions called lobules, each of which is 1.20th of an inch in diameter, and composed of little branches of the blood vessels, and of the liver duct, the interstices being filled with liver cells. The latter form the secreting part of the gland. They are spheroidal, and contain little nuceli and granules. Sometimes they exhibit slow contractile movements, just as those little animals which are termed amœbæ do, or the white blood corpuscles of the blood which are very similar to amœbæ, and, like them, throw out arm-like prolongations, and then withdraw them again.

Disease of the liver is very rare in the horse in this country, but is more often met with in eastern countries, especially in India. The usual causes are high feeding, and want of exercise ; while residence in hot, damp climes also predisposes horses to attack. Lack of sufficient air is also regarded as a cause of this affection. The usual signs of liver disease are the same in the horse as in man. Among the most common symptoms presented, are jaundice, local pain in the region of the liver, colicky pains, and persistent pain in the off shoulder. From a comparative point of view, it is interesting to observe how these manifestations of liver disease, correspond with those shown in man, in whom the lameness of the horse's *right fore-limb* is represented by pain in the *right shoulder*.

In cases of congestion of the liver coming on *suddenly*, and manifested by jaundice, shown by the yellowish tinge of the white of the eyes, and loss of appetite, coldness of the extremities, and pain over the region of the liver on pressure, it is advisable to abstract blood in moderation, say two or three quarts. The blood-letting should be followed up by the administration of five or six drachms of aloes. The diet should be laxative and restricted in amount. After the aloes is given, sulphate of sodium in four ounce doses with one ounce of bicarbonate of potassium may be given twice daily in the drinking water. When congestion is *gradually* developed, as it sometimes is in hunters "summering," and is dependent on dietic errors or want of proper exercise, it is of the first importance to restrict the amount of food, and attend to the sanitary conditions. In addition, salines, such as those above recommended, may be administered twice daily in the drinking water, and a full dose of aloes given.

In order to prevent the recurrence of these affections of the liver, the diet should be restricted in amount, and regular exercise enjoined.

Experience has abundantly proved that a hunter is all the better for his winter work, if "summered" on his ordinary food. The food, of course, should be given in *smaller quantities*, and the horse should be kept *in regular, though not exacting exercise.*

CHAPTER IV.

INTESTINAL PARASITES.

Bots. Ascaris Megalocepha, or Large-headed Lumbricoid Worm. Oxyuris Curvula, or Maw Worm. Strongylus Tetracanthus, or Four-Spined Strongyle. Strongylus Armatus, or Armed Strongyle. Echinococcus Veterinorum, or Common Hydatid of the Horse.

ALTHOUGH one is generally disposed to hear and speak of worms with anything but pleasurable feelings, we need hardly tell our readers that there are few creatures, whose development and growth are more wondrous or more interesting, than that of the several intestinal parasites. It is not, however, our purpose to enter deeply into the life histories of these creatures, but to cast a glance over the modes of their living, and to describe in as simple a manner as we are able, the methods to be adopted in order to rid the horse of his self-invited guests. We shall treat of six different varieties of internal parasites. The other kinds are so rarely met with, as to require no description at our hands.

The first of the parasites of the horse of which we have to speak is the common gad-fly, or œstrus equi, whose larval form is the bot. All our readers have heard of the gad-flies, which prove so irritating to oxen by piercing through their hides. The female gad-fly settles on its victim while out at grass, late in the summer, not for the sake of deriving sustenance for herself, but for that of providing a suitable habitat for her eggs. It is at this time of the year that she deposits her eggs on the hairs of the coat, and this she is enabled to do by means of a thick, sticky fluid. The fly generally selects, as sites for depositing her eggs, those parts of the horse which the animal can reach easily with the tongue, namely the shoulders, the lower part of the neck, and the inner parts of the forelegs, especially around the knees. The horse frequently licks the portion of coat on which the eggs have been deposited. They gradually become hatched in about three weeks from the time of their deposition by the gad-fly, and the larval form or maggot makes its escape out of its enclosing egg-shell. The maggots are then carried to the horse's mouth, and ultimately to his stomach along with his food and drink. Necessarily, as Professor Williams points out, many larvæ perish during this passive mode of immigration ; some being dropped from the mouth, and others being crushed in the food during mastication ; but notwithstanding the waste, the interior of the horse's stomach may become completely covered with the larvæ commonly termed "bots."

When the bots, which hold on to the lining of the stomach by means of two large hooks, are perfectly grown, they release themselves, and are carried through the intestines along with the excreta, and thus they finally fall to the ground. They then bury themselves below the surface, in order to undergo a transformation from the condition of the bot, to that of the pupa or chrysalis. When they have remained thus buried for six weeks, they make their way out of their enclosing cocoons, and emerge as perfect gad-flies. The male insects die, but the females live long enough to deposit their eggs, which are generally about forty in number. The bot passes about eight months of its existence in the stomach, where it is present in the winter months. It leaves in spring or early summer. The fly is developed from June to September, and after the latest females have appeared, all perish in October. Not uncommonly, the presence of bots in the stomach of the horse gives rise to considerable mischief. When very numerous, they may set up serious disease, sometimes even perforating the walls of the intestines. It is not always possible to diagnose the presence of bots in the stomach, but not unfrequently they may be seen in the excrement, or adhering around the anus. When there is reason to suspect their presence in very large numbers, it is well to place a piece of rock salt in the horse's manger, and to administer a drench, consisting of spirit of ether two ounces, of glycerine of carbolic acid three drachms, and of linseed oil a pint. This may be repeated once every day for four days.

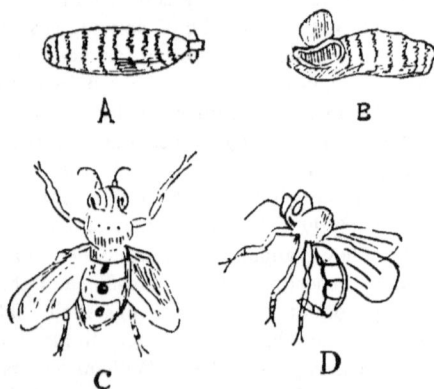

A   B

C   D

The accompanying pictures show the various forms assumed by the œstrus equi. A is the larva or bot, B is the pupa case, C is the male fly, and D is the female fly. There is another species of gad-fly called the œstrus hæmorrhoidalis, which deposits its eggs on the lips and nostrils of the horse. We need not speak at length here of the bot-fly of the ox, but may mention that, unlike the bot-fly of the horse, it passes its larval stage as a bot beneath the skin of its host, and it is this larva whose growth causes the appearance of the tumours called warbles. The fly is provided with an ovi-positor, by means of which it bores holes through the skin, in each of which it deposits one egg. The eggs develop into bots, which may be recognised by the growth of little elevations or tumours, called warbles. The tail end of the bot places itself in the tumour of the host, in order that it may be enabled to breathe. The bot, when completely developed, escapes and buries itself in the ground, and then passes through the chrysalis stage, in which it remains about six weeks, at the end of which time it at length emerges as a perfect fly, which again deposits eggs, and so the cycle goes on again. The eggs of the bot-fly of the sheep are deposited in the nostrils, to the great distress of the poor animal. The larvæ or bots pass upwards towards the sinuses or cavities of the forehead of the sheep, and thus cause great distress.

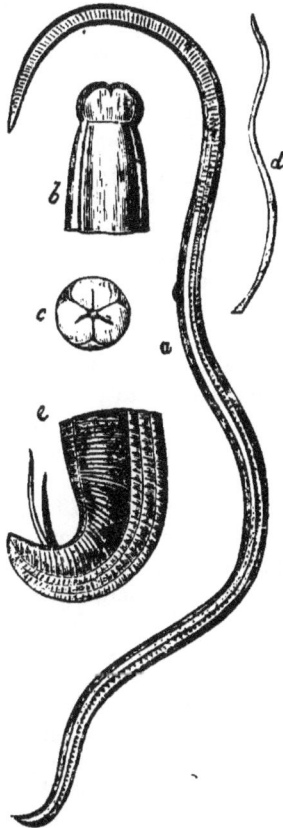

ASCARIS LUMBRICOIDES.

A, Female Ascaris Lumbricoides. B, Anterior extremity enlarged, seen from the side.
C, The same, seen from the front, showing the opening in the centre.
E, The Posterior extremity enlarged.   D, Male Ascaris, natural size.
G

The worms which invade the horse most commonly, are the ascaris megalocephala, or the large-headed lumbricoid worm, and the little oxyuris curvula or maw worm. The male of the lumbricoid worm is rarely over seven inches, but the female may attain a length of sixteen to eighteen inches. This worm has a smooth body with transverse rings, and it may occur in any part of the intestinal canal, although it is especially found in the small intestines. The horse is known to be invaded by these parasites by their occasional passage out of the body with the excrement.

It is noteworthy that the eggs of the lumbricoids effectually resist dryness, and it is possible that horses become infected with this parasite, by drinking out of ponds containing sewage matter. In cases of invasion by this parasite, the horse should have a full and nutritious diet. In the first instance, it is well to give a ball containing six drachms of aloes and one drachm of tartar emetic. Two or three days afterwards, a ball consisting of santonine thirty grains, of sulphate of iron one drachm, of carbolic acid fifteen drops, of aloes one drachm, made up with ginger and gentian and treacle to one ounce, may be given twice daily for three or four days.

The oxyuris curvula, or maw worm, is partly transparent, and is marked with transverse stripes. In length, the males are one and three quarters of an inch, and the females from three to four inches or more. This worm has a long tail. Its usual habitat is in the large sacculated bowel, where it may set up much local irritation.

The horse rarely requires treatment for the expulsion of this worm. A full dose of aloes may be administered, and clysters given once daily for a few days will prove beneficial. Each clyster may be made of two ounces of the oil of turpentine, with mucilage of starch one pint.

We will now turn our attention to two other kinds of worms which not infrequently infest the horse. One is called the four-spined strongyle (Strongylus Tetracanthus), the other the armed strongyle (Strongylus Armatus).

The four-spined strongyle is sometimes found in large numbers in the horse, ass, and mule, and, though not uncommon in this country, is not so frequently met with on the continent. The males and females are of about equal size, and occupy the walls of the large intestines of their hosts. They set up by their presence localised congestion and inflammation, and the formation of matter in the wall of the gut. The species is recognised by its bright red colour, by the four conical spines surrounding the mouth, by two neck bristles, and by the three-lobed long head of the male. The head, when viewed in profile, is truncated, and seen from above it appears round. The body is smooth, and presents indistinct rings. The eggs of the worm probably gain access to the intestines of small insects, and the immature form is swallowed by the horse in the water or in the food. The worms then become encapsuled in the lining membrane of the large gut. They are then about 3·6 millimetres long when uncoiled.

In this condition the worms cast their skins. They enter the cavity of the gut, and undergo another change of skin prior to acquiring the adult state. They do this by rolling themselves within the fæcal matter of the horse's intestine. In this state they lie coiled in the cocoons they make for themselves. In some districts the worm is not often met with, while in others it is most destructive. Mr. Lloyd was the first to recognise this worm as the cause of the Welsh epizootic outbreaks. The worm gives rise to emaciation, colic, diarrhœa, and sometimes to inflammation of the bowels. Sometimes the pain is very acute, and the animal rolls and tosses about in great agony. In other instances, abdominal pain is not a marked feature. Last year we were called to see a team of cart horses, two of which had already died. The remaining two were much prostrated and extremely emaciated. Careful examination of the excreta soon revealed to us these characteristic little red worms, as the cause of the mischief. The males are about one-eighth of an inch long, the females two-fifths of an inch. When these worms infest the horse, a full dose of aloes should be given in the first instance, and this should be followed up by giving mashes for a couple of days or so, and then by good nutritious food and the administration of tonic and stimulating medicines. The following formula is a good one :--of carbonate of ammonium two drachms, the double citrate of iron and ammonium two drachms, ginger, gentian, and treacle, to make an eight drachm ball. One ball to be given two or three times daily for three weeks.

We may now say a few words regarding the armed strongyle. This worm has long been known to naturalists. Formerly two varieties were described, but these are now known to be the same worm in different stages of growth. The body is rigid, the head flattened and armed with numerous upright denticles like those of a circular saw. The hind ray of the hood of the male is thrice cleft. The males are about an inch and a half long, the females two inches. The eggs are elliptical, and when passed out with the fæces, they become hatched in three weeks in mud, and at the same time part with their tails. They next gain access to the bodies of some intermediary host, probably some insect, where they are still further matured, and from thence they gain access to the horse. From the intestinal

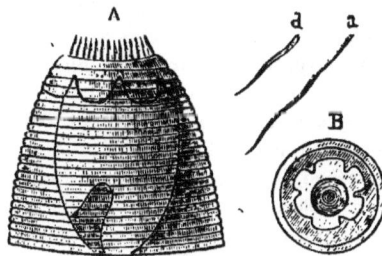

ARMED STRONGYLE.

a, adult strongyle, natural size ; A, head of adult, enlarged thirty times. d, asexual strongyle, natural size ; B, head of asexual strongyle, seen from the point, enlarged thirty times. (Zundel).

canal of the horse they get into the small vessels of the gut, and pass on into
the large arteries, and becoming embedded in the walls of these larger blood
vessels, they cause bulgings termed aneurisms. Lastly, these parasites
make their way out of the vessels, and thence endeavour to make for the
large gut, where they again change their skins, and, adhering to the mucous
lining of the bowel by means of their spines, attain sexual maturity.

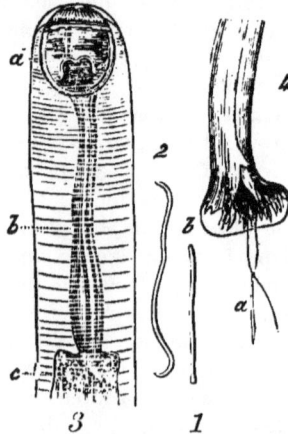

ANEURYSMAL ARMED STRONGYLE.

1, male, natural size; 2, female, natural size; 3, anterior extremity, highly
magnified; a, complete buccal capsule; b, œsphagus, or gullet; c, intestine.
4, caudal extremity of the male worm; A, hook and accessory
part. (Rayer).

The aneurysms or swellings occur in a large percentage of horses and
asses, and they vary in size from a pea to a man's head, and are met with in
animals six months old and upwards. The number of worms in an aneurism
varies, and is usually nine to ten, the highest number in one horse being 121.
It has been said "that foals and yearlings suffer more from parasites in the
paddocks, than they do on adjoining farms where only a few animals are
bred." "This is explained," says Dr. Cobbold, "by the relatively greater
amount of egg dispersion proceeding from the infected brood mares. It is
quite evident that the lives of many valuable animals are annually sacrificed
by the neglect of hygienic arrangements. The palisade worm, as this
parasite is commonly called, is chiefly destructive to young animals, and, as
Mr. Percivall has remarked, is commonly the cause of lingering and hidden
disease terminating in death, without any suspicion on the part of the
practitioner as to the nature of the malady." Treatment for these worms is
not of much value.

We have lastly to speak of the common hydatid of the horse. It is well
known that after death hydatids *(Echinococci veterinorum)* are
sometimes found in the various organs and glands of the horse, more
especially in the lungs, liver, and kidneys, and sometimes in the brain.

They vary much in size, sometimes being as small as a pea, and occasionally as large as a good-sized cocoa-nut. They may or may not produce symptoms, which vary according to the organ invaded, and the size and exact position of the cyst. This hydatid is common in man, being often found in the liver, and sometimes in other organs ; and it may attain in him a very large size. This hydatid or cyst in its early form is small and globular, with transparent walls and finely granular contents. In its later stages, when it has much increased in size, the walls become thick, and the contents fluid. Sometimes these cysts contain several pints of clear fluid.

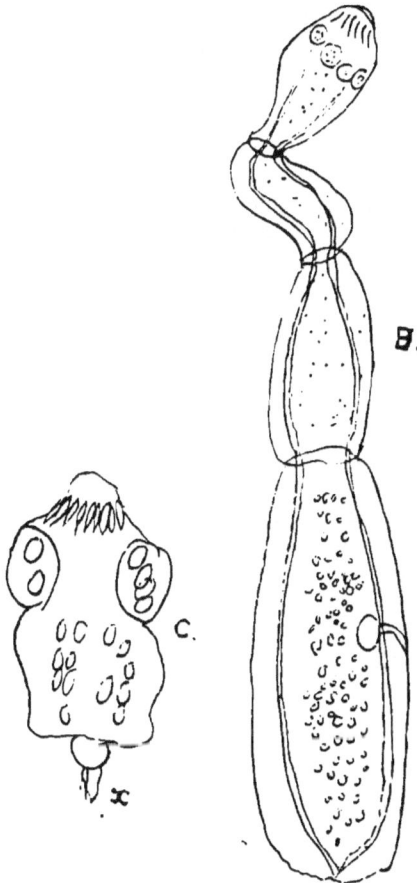

In the above picture B shows the echinococcus of the dog, magnified ; C is one of the little heads which are formed in the cyst wall ; X is the part where the head is attached to the cyst wall.

This hydatid is the larval form of the *Tænia Echinococcus*, which infests the small intestines of the dog or wolf. The adult tape worm is composed of four segments or joints, and is a little over a quarter of an inch in length. The first joint includes the head, which is about one-hundreth part of an inch wide, and is provided with four suckers, a double coronet of hooklets, between thirty and forty in number, and a central beak. The fourth segment is as long as all the rest of the worm. The way in which the human being and the horse and other creatures become infested with this larval form is as follows :— The mature worm in the intestines of the dog discharges its ripe eggs, and these being ingested by man or the horse, soon lose their shells, which are dissolved, thus liberating the six-hooked little embryoes. These bore their way into one of the blood vessels, and are thus carried to the various organs of the body, more especially the liver and lungs. When the embryoes have arrived here, they become metamorphosed into hydatids. The lining membrane of these little cysts then develops heads. The worm cannot undergo further development, unless the hydatid be eaten by some animal.

Dr. Cobbold tells us that at least 1 per cent. of our dogs harbour the mature tape worm, and he asserts that in the United Kingdom several hundred human deaths occur annually from the ingestion of the eggs, which develop into hydatids. In some other countries, especially in Iceland, where dogs are so much used, and live in close contact with their masters, this disorder is fatally endemic, and thus Iceland stands at the head of the afflicted territories. Our Australian colonies are probably entitled to the next place of distinction in this respect (Cobbold). Dogs frequently convey the eggs of this parasite to man by licking his hands and face. Regarding the treatment of this larval form we have nothing to say, it being very rarely diagnosed in horses. Last year we had under treatment an aged cart horse suffering from chronic renal disease. The water passed contained abundance of matter. After death, thirty hydatid cysts were found in the right kidney. They varied considerably in size, one being as large as a cocoa-nut ; the others varied from the size of a walnut to that of a pea. In the left kidney there were also more than a dozen of these cysts.

CHAPTER V.

DISEASES OF THE KIDNEYS AND BLADDER.

*Inflammation of the Kidneys.   Retention of Urine.   Incontinence of Urine.
Stone or Calculus in the Bladder.   Inflammation of the Bladder.*

THE kidneys are two glands whose chief function it is to eliminate from
the blood certain substances, the products of the waste of the various parts
of the body.   They vary much in weight in different horses, but the right
one is always more voluminous and heavier than the left, its average weight
being twenty-seven ounces, while that of the latter is only twenty-five ounces.
The diseases of the kidneys and bladder in the horse are not nearly so
frequent or so varied as in man ; but nevertheless they merit careful
attention, for interference with the functions of these intricate glands is of
serious moment.

Before describing the diseases of the kidneys, we may say a few words
regarding the conformation and structure of these important organs.   In the
horse, sheep, and pig, the kidneys are not composed of distinct lobules as
they are in the ox ; although during development they present a similar
conformation.   If the kidney be carefully examined with the microscope,
it will be found to consist of a large number of tubes, made up of several
distinct sections, which differ very much both in situation and in structure.
Anyone who has not made a special study of the wonderful conformation of
these little organs, would hardly credit the wondrous formation, and the
labyrinthiform intricacy of their secreting conduits, lined with variously
shaped cells.

The little tubes or conduits commence as dilated capsules, composed of
fibrous tissue, and are lined internally with little flattened plates called
epithelium cells.   Inside the capsule, will be seen a tuft of very small blood
vessels bound together by tissue, and likewise covered by flattened epithelial
plates.   The tuft of vessels has a main vessel leading to it, and one leading
from it.   The blood brought by the former is freed from water and salts in
the capsule, and it returns purified through the latter.

A is the capsule.   B is the tuft of vessels.   C and D are the two vessels,
of which one enters and one leaves the tuft.   E is the commencing tube
lined by cells.   The tubes are on an average about one six-hundredth of an
inch in diameter, and as they pass onwards, they vary greatly in shape, and

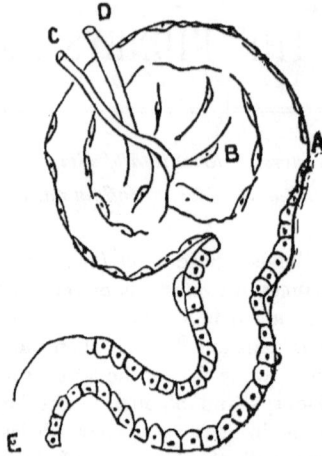

are lined by special cells, whose duty it is to separate the waste products in
the blood. ¿The accompanying diagram shows the varying contour of the
tubes in the various parts.

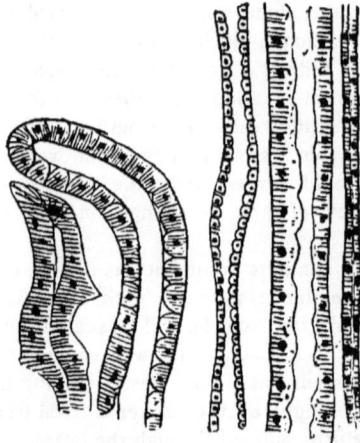

After these preliminary remarks, we may immediately proceed to
consider the various derangements of the kidneys and bladder of the horse.
The first disorder to which we propose to draw attention, is acute
inflammation of the kidneys. This is a disease for the most part due to
chill or exhaustion. It is attended by considerable fever with colicky pains,
the attack resembling colic attended by fever. The pulse is quickened, and
is full, hard, and firm. The breathing is short and accelerated, the bowels
are constipated, and there is much thirst. In some instances there is
stiffness, tenderness in the loins, and arching of the back, but these
symptoms are not invariably present. The most characteristic feature of

inflammation of the kidneys, however, is the scanty elimination or total suppression of the urine, and the desire to pass it frequently. The animal strains violently, but may be unable to pass more than a few drops of water. This is highly coloured, and contains blood. In many cases of inflammation of the kidneys, the flow of water completely ceases ; and not uncommonly there is no other symptom to indicate the nature of the malady. In other cases, there are signs of pain manifested by lying down and rolling about ; while at other times the seat of pain is pointed at, by the animal turning round, and endeavouring to bite or scratch at the loins. If the suppression of the urine be prolonged, the animal may become partially unconscious ; but there is no loss of motor power.

In most cases of acute inflammation caused by cold, fatigue, or exhaustion, if the pulse be strong, it is our custom to bleed in moderation, that is to remove from two to three quarts. Those cases, however, which supervene on various fevers, do not bear depletion so well. The bowels must be freely acted upon by the administration of five to six or seven drachms of aloes. A lax condition of the bowels should be maintained by the administration of sulphate of sodium given in the drinking water. Eight to twelve ounces or more may be given in the course of the day. The pain may be relieved by the application of woollen cloths, which have been steeped in hot water and then wrung out, or of linseed-meal poultices over the loins. The diet should be laxative, consisting of linseed and oatmeal gruel and bran mashes. When the fever has subsided, salts of iron and vegetable tonics are necessary. The following formula is a good one for this and other cases of horses convalescing from acute inflammations :—Of sulphate of cinchonine, forty grains ; of the double citrate of iron and ammonium, two drachms, made up into a ball with gentian, ginger, and treacle to eight drachms. One of these balls may be given twice daily for four or five days.

We may now speak of the presence of blood in the urine. One of the causes of this occurrence, as we have just said, is inflammation of the kidneys. There are, however, other causes of this condition, such as strains from violent exertion, improper feeding, diseases of the bladder such as inflammation, of which we shall shortly speak, and the presence of a stone in the bladder. In these cases, the animal must be put on a plain, laxative, soothing diet, which should consist of linseed gruel and mashes. Linseed tea is a very good drink in the place of water. Three drachms of aloes, in the form of a ball or in solution, may be given in the first instance ; and in the drinking water three drachms of tincture of perchloride of iron may be given twice daily, so long as the condition of the urine remains unaltered. Sometimes a condition is met with in which the urine is very high coloured, and yet does not contain blood. For this, luxurious dieting is to be assigned as the cause. In such cases, five or six drachms of aloes may be administered in the first instance, and half an ounce of bicarbonate of potassium may be given twice daily in the drinking water. We have already treated of diabetes or profuse urination, which, as we mentioned, is not a disease of the kidneys.

Retention of the urine is a condition not very uncommon in the horse, and is dependent on a variety of causes. The animal, although attempting to pass water, is unable to do so. In these cases the bladder becomes much distended, and the animal stretches himself and strains violently, sometimes groaning with pain. In these cases a clyster of warm water, in which four drachms of opium have been boiled, has been recommended. The animal should be warmly clad, and a mild dose of physic, such as three-quarters of a pint of linseed oil, may be given. A ball consisting of camphor two drachms, and of opium one drachm, has proved useful, according to some authorities, when administered early, and repeated in a hour or two. When these measures are ineffectual, it will be necessary for the veterinary surgeon to pass the catheter.

Incontinence of urine is a condition likewise depending upon several different causes. In some instances retention is accompanied by incontinence of urine, which continually dribbles away. In other cases it is due to stone in the bladder, or to paralysis of the orifice of this organ. When incontinence is due to over-distension, the catheter must be used. When due to paralysis, clysters of cold water into the rectum, and the administration internally of balls consisting of powdered nux vomica a drachm, and of ginger and gentian with treacle to eight drachms, may be given twice or thrice daily.

We have lastly to speak of stone in the bladder, and of inflammation of this organ. A calculus or stone is composed of varying substances, but in most cases contains a large quantity of carbonate of lime. It varies much in size and consistency, and may sometimes almost fill the cavity of the bladder. Stone is usually manifested by repeated straining, and attempts to pass water, colicky pains, incontinence of urine, repeated motions of the tail, stiffness of the hind limbs, and by interruption to the flow of urine. Surgical interference is the only treatment of any avail in cases of stone. The accompanying picture is a section of a calculus from the bladder of the horse, showing the disposition of its constituents in concentric circles.

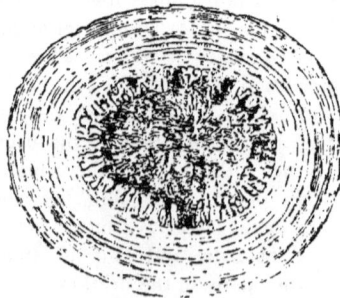

Inflammation of the bladder in the horse is nearly always caused by the administration of cantharides or turpentine, or by the absorption of cantharidine from a large . blister of cantharides. This malady may however, also be set up by the presence of a stone in the bladder. The symptoms of inflammation of this organ are restlessness, pain, and frequent attempts to pass water, which are attended with difficulty and pain. Febrile symptoms are also present in most instances. The nature of the case is known by the history, if it occurs after severe blistering. In such cases the blister must be at once rubbed off. The bowels should be regulated by laxative diet, and demulcent liquids should be allowed. Hot fomentations applied to the abdomen are very beneficial in alleviating the pain when severe. Internally, drenches composed of ten minims of Fleming's tincture of aconite, and four ounces of liquor ammonii acetatis may be given every six hours for the first day. Afterwards the liquor ammonii acetatis may be given alone, three times daily.

We may conclude our notice of the diseases of the kidneys and bladder, by adding that in those cases, where, from a variety of causes, the water is observed not to be passed as readily as it should be, half an ounce of bicarbonate of potassium, with one drachm of nitre, may be given in the drinking water, or with the food, twice daily.

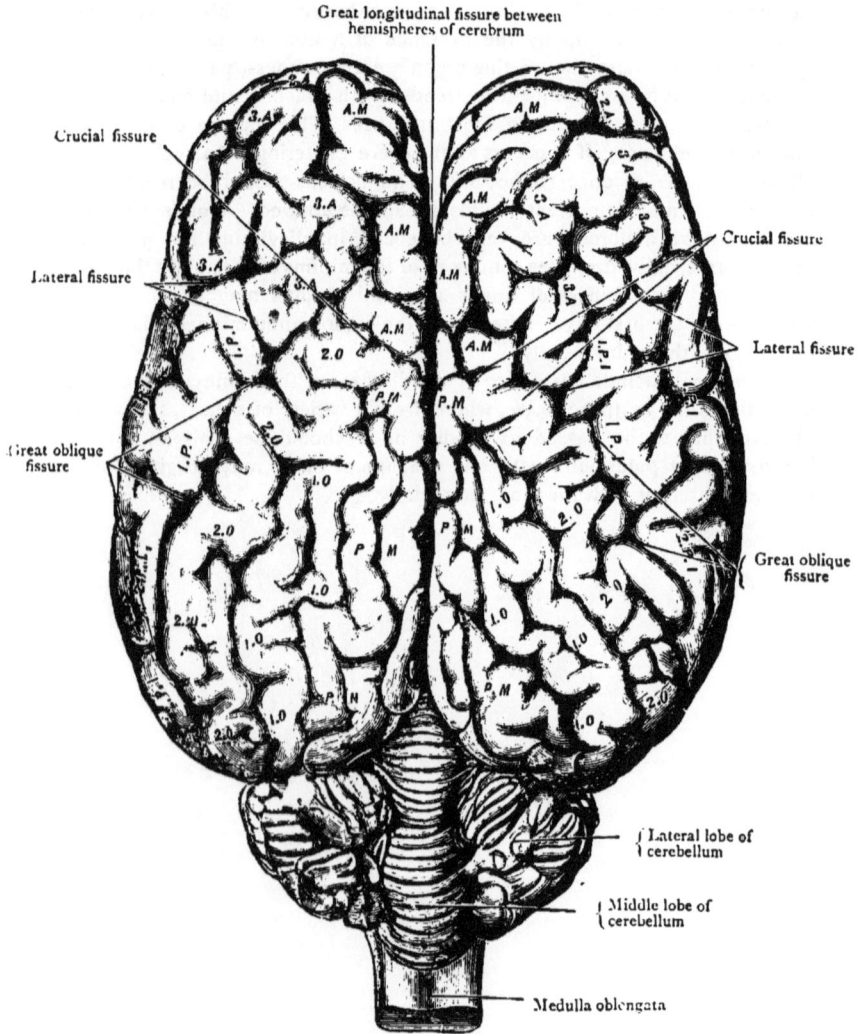

Great longitudinal fissure between
hemispheres of cerebrum

Crucial fissure

Lateral fissure

Great oblique
fissure

Crucial fissure

Lateral fissure

Great oblique
fissure

{ Lateral lobe of
{ cerebellum

{ Middle lobe of
{ cerebellum

Medulla oblongata

BRAIN—SUPERIOR ASPECT.

Great longitudinal fissure between
hemispheres of cerebrum
Olfactory bulb
Olfactory Peduncle
Int. olf. tract
Infundibulum
Tuber cinereum
Optic (2nd) nerve
Optic chiasma
Optic tract
Pituitary body
Fissure of } Sylvius }
Corpus } albicans }
Ext. olf. tract
Pons } Tarini }
3rd nerve
4th nerve
int. root } Ext. root } of 5th nerve
6th nerve
7th nerve
Portio } intermedia
Crus cerebri
8th nerve
Great oblique } fissure }
9th nerve
Pons Varolii
10th nerve
Trapezium
11th nerve
Lateral lobe } of cerebellum }
12th nerve
Inf. pyramid
Medulla oblongata
Decussation of pyramids

BRAIN—INFERIOR ASPECT.

CHAPTER VI.

DISEASES OF THE NERVOUS SYSTEM.

General remarks on the Anatomy and Physiology of the Nervous System of the Horse. Stringhalt, Chorea, Shivering, "Immobilité." Megrims, or Congestion of the Brain. Mad Staggers, or Inflammation of the Brain. Epilepsy, Paralysis, Hydrocephalus, or Water on the Brain. Tumours of the Brain.

GENERAL REMARKS ON THE ANATOMY AND PHYSIOLOGY OF THE NERVOUS SYSTEM.

DISEASES of the nervous system, as might be expected, are not nearly so frequently met with in the equine tribe as they are in man, and they present far less diversity of form and character. It is sufficiently clear that amid all the marchings and counter-marchings which have been taking place in the rapidly advancing civilisation of man, the most forced and rapid advancement is that which has been aptly termed the "march of intellect." But, like other forced movements, it has been attended by many heavy penalties ; for all forced marches, when repeated frequently, wear out the finest troops that were ever commanded by energetic generals. So has it been with modern intellectual advancement, rendered imperative by the growing demand of progressive civilisation, which has been attended, as has been known for some time past, by those many forms of nervous exhaustion, which appear before us like spectres at every step.

It is not our purpose here, to discuss this part of our subject further, but we may point out that the horse, by nature retiring, timid, and excitable, although as far as we know free from nervous diseases, while enjoying liberty, untouched by the hand of man, has likewise become subject to a list of maladies, fortunately not a long one, the results of confinement, and the artificial conditions which attend it. We shall treat of stringhalt, "shivering," chorea, megrims, mad staggers, epilepsy, paralysis, water and tumours in the brain. Before describing these maladies, we may first consider briefly a few of the most important structural features of the nervous system ; for these are of very great interest and importance. The nervous system of man and the higher animals consists of two portions—the cerebro-spinal and the sympathetic—each of which has certain characteristics in structural build, in range of influence, and in mode of action. The cerebro-spinal system includes the brain and spinal cord, and the various nerves proceeding from them. The sympathetic system consists of a double chain of nerves—one on each side of the backbone—from which branches are distributed. The nervous apparatus is made up of two ultimate factors, nerve-fibres and nerve-cells, and these are intimately associated together. The cells are collected together in groups or masses, and are always mingled more or less with fibres, and both together form what is termed a "nerve-centre." The fibres, besides entering into the composition of nerve-centres, form nerves, which connect the different centres, and are distributed to the various parts of the individual. Nerve-cells and nerve-fibres differ in function. The former generate and conduct nerve force, while the latter merely conduct it. We may compare the nervous system with a galvanic battery, and the telegraphic wires proceeding from it. The battery, like the nerve centre, generates and conducts the current; while the wires, like the nerves, merely conduct it, having no share whatever in its production.

Nervous force travels at a very quick rate. It has been calculated by physiologists, that the rate of conduction in human nerves supplying muscles with motor power is 111 feet per second, and that in those nerves by which sensation is conducted, it is still quicker, reaching as high as 140 feet per second. Each nerve is composed of a variable number of bundles of nerve-fibres, which have separate sheaths. The bundles of fibres, also, have separate sheaths, and the whole of them in turn are enclosed in a firm fibrous covering.

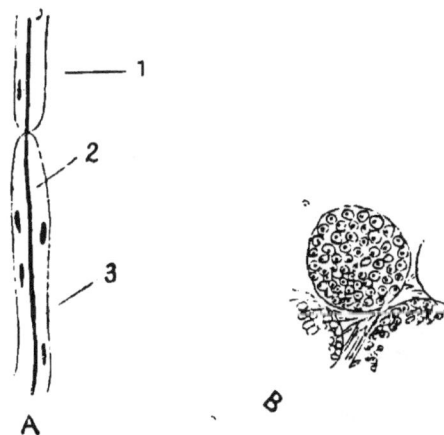

Figure A shows a nerve fibre (after Klein) magnified 300 times. 1 is the sheath, 2 is the medulla, and 3 is the axis-cylinder. The constriction in the centre, where the medulla is deficient, is called a node of Ranvier. Figure B shows a bundle of nerve-fibrils cut transversely, and parts of two others. Several such bundles make up a nerve. This specimen is from the nerve of a dog, highly magnified.

A nerve-fibre is a microscopic element composed of a proper wall and contents. The wall is the sheath we mentioned, and it is a thin elastic membrane. The contents comprise in the centre a solid core, called the axis cylinder, along which the nerve current passes. In many fibres, between the axis cylinder and the wall, is found a viscid substance called the medulla. Those fibres which do not contain the medulla, and which are specially characteristic of the sympathetic system, are called non-medullated.

The majority of nerve-fibres measure about $\frac{1}{2800}$ of an inch in diameter. The nerve-cells are large nucleated bodies of very variable shape, and they have one or more prolongations extending from them. These prolongations or poles establish relations with the nerve-fibres, and constitute the origin of the nerves.

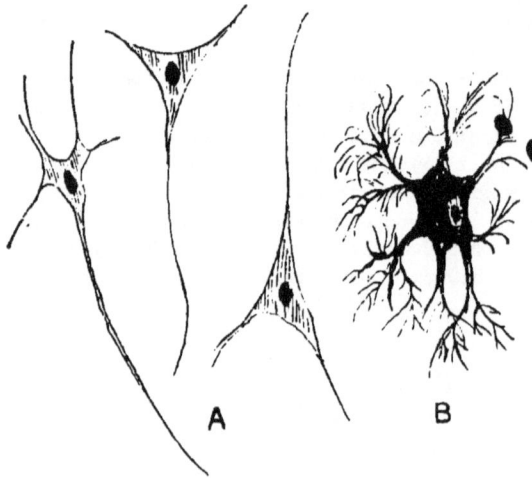

In the above figures, A shows some nerve cells of different shapes. B shows a stellate cell from a developing animal, magnified 400 diameters. When a cerebro-spinal nerve is irritated by pinching, there is either pain manifested, or there is twitching of one or more muscles, to which the nerve distributes its fibres. From various considerations, it is certain that pain is always the result of change in the nerve cells of the brain. Therefore, in such experiments as those referred to, it seems to the experimenter that the irritation of the nerve-fibre is conducted in one of two directions, either to the brain the central termination of the fibre, when there is pain, or to a muscle when there is movement. The effects of these simple experiments are the types of what always occur, when nerve-fibres are engaged in the performance of their functions (Kirke). The brain of the horse and of the other higher animals is formed of a central white part composed of fibres, and an outer convoluted portion of grey matter composed of nerve-cells and fibres. In the horse it weighs about 23 ounces, in the ass 12 ounces; and it is formed of a front portion called the cerebrum, and a hind part called the cerebellum. The spinal cord, unlike the brain, is formed of an outer white portion and a central grey portion, the former made up of fibres, and the latter of cells and fibres.

The two figures placed at the beginning of this chapter are taken by the kind permission of Dr. J. McFadyean, M.B., C.M., B.Sc., from his valuable work on the "Anatomy of the Horse." They will give our readers a very good idea of the superior and inferior aspects of the horse's brain. It will be seen that the encephalon, or brain of the horse, is an ovoid mass, which, when viewed on its superior surface, shews most posteriorly the continuation of the spinal cord, called the medulla oblongata, and in front of this, the superior surfaces of the middle and two lateral lobes of the cerebellum. In front of the cerebellum are seen the two large cerebral hemispheres, which are separated from the cerebellum by a deep transverse fissure, into which the tentorium cerebelli passes.

On the inferior aspect, we see that the medulla oblongata is prolonged beneath the cerebellum, and then becomes continuous with the cerebral hemispheres, by means of the crura cerbri, which are bounded in front by the two thick white cords, the optic nerves. The brain may be said to consist of three portions : (1) *The isthmus of the encephalon* (the prolongation of the spinal cord); (2) The cerebellum ; (3) The cerebrum.

For further details, vide Dr. McFadyean's work, or Chauveau's Comparative Anatomy.

STRINGHALT.

AFTER these preliminary remarks, we may at once proceed to describe the diseases of the nervous system of the horse, commencing with stringhalt. Stringhalt consists in involuntary convulsive motions of the muscles, generally those of one or both hind legs ; but occasionally it is seen in the fore legs also. Generally speaking, however, it is confined to one of the hind legs ; more rarely affecting both of them. Stringhalt is a common affection of the horse, and of necessity constitutes unsoundness, although many horses affected with this disorder are able to do their work exceedingly well. We have a chestnut horse at the present time, and have seen numbers of others, which do their work every whit as well as horses in all respects healthy. In severe cases, stringhalt is evident to the observer at every step taken by the animal, while in cases not so marked, the affection can only be noticed at longer or shorter intervals. The animal may proceed a few yards in a normal manner, and then suddenly snatch one or both of his hind legs from the ground convulsively, with a sudden jerk, and bring it down again with unusual force.

Stringhalt often becomes worse as time passes on, but it may remain in pretty much the same condition for some years. We have often observed that it improves as the general health and condition of the animal improve, and becomes worse when the animal is worked too hard, or when from any other cause he is out of condition. As, in many instances, stringhalt

H

becomes more aggravated with age, and as the value of animals afflicted with it, is depreciated by this unsoundness, it is important to be able to recognise the affection, when only slightly developed. It is advisable to have the animal turned from one side to the other, and then in the reverse direction. In cases of slight stringhalt, the peculiar convulsive twitching is often shown only as the animal turns one way.

Stringhalt is a disease which generally comes on gradually, but cases where the malady has come on in the night are recorded. We are of the same opinion as Professor Williams, in considering that the chief cause of stringhalt is an inflamed condition of the nerves, supplying the affected limb. Stringhalt coming on more rapidly, is in many cases a rheumatic affection, due to cold, damp, or exposure.

Although on the continent, methods of treating chronic stringhalt by certain surgical operations have been advocated, yet at present we are not able to say how far these methods have been successful. This subject, indeed, is at present engaging the attention of veterinarians at home and abroad. In cases where the symptoms become aggravated from any cause, or when the disease suddenly manifests itself, the animal should be rested ; and if the disease be due to rheumatism, the malady should be treated as we have already mentioned. In such cases nothing need be done locally, beyond hot water fomentations. When not traceable to rheumatism, rest, a dose of physic, hot fomentations, and three drachms of bromide of potassium, given in the drinking water three times daily, may prove serviceable.

CHOREA, OR ST. VITUS' DANCE.

WE will now speak briefly of the disease called chorea, and better known in the human being under the name of St. Vitus' dance. It is a peculiar disorder, characterised by irregular contractions of different muscles. It is not a common malady in the horse. It is usually traceable to hereditary predisposition, although mal-hygienic conditions, overwork, and exhaustion, may also act as exciting causes. Stringhalt itself may be regarded as a peculiar choreic disease. In treating chorea, it is necessary whenever it is possible, to remove the cause when that is to be ascertained. The general hygienic conditions should be attended to, the diet should be good and nutritious, and the work *proportioned* to the *strength* of the animal. Internally half an ounce of Fowler's solution, and three drachms of bromide of potassium, may be given twice daily in the drinking water. "Shivering" is a peculiar disorder affecting the muscles of the back and posterior extremities. When a horse subject to this affection is backed or turned, the muscles of this region are thrown into a spasmodic condition, contracting and relaxing irregularly. The tail is often spasmodically elevated, and then depressed. When the horse is trotted forwards, the spasms are very seldom developed, but they may be brought into action by the head being rapidly turned round.

SHIVERING, "IMMOBILITÉ."

" Shivering " is so called from the resemblance of the muscular spasms to shiverings. *Immobilité* is the word which the French apply to those cases of muscular weakness which are manifested by the inability of the horse to turn round quickly, without falling. The horse can walk or trot forwards, but when turned sharply, he falls to the ground. Sometimes a horse turns with great difficulty, but does not actually fall. He moves his hind limbs in an unsteady and irregular way, and seems to have but little power of co-ordinating the movement of this part of the body. This latter condition* is generally termed by horsemen broken or sprained back, and is usually due to chronic disease of the spinal marrow. Shivering, *immobilité*, and sprained back all constitute unsoundness. Professor Williams records that four young horses, the progeny of a dam which was affected in the back, died from paralysis of the spine, before they had attained the age of three years. A fifth is now living, and shows signs of aggravated nerve disease. Nerve disease is commonly transmitted to the offspring in the equine tribe, and we mentioned in treating of roaring, a nervous disease often dependent upon dietetic mismanagement, the important part which hereditary disease plays in the production of this malady.

MEGRIMS, OR CONGESTION OF THE BRAIN.

We turn now to the remaining disorders of the nervous system of the horse, namely, megrims, mad staggers, epilepsy, paralysis of the lips, water and tumours in the brain, and lock-jaw or tetanus. We have, in treating of diseases of the stomach, spoken fully of stomach staggers, and in treating of poisons, we spoke of grass staggers. Now, we have first to consider the two remaining varieties of staggers, and these are megrims or congestion of the brain, and mad staggers, or inflammation of the brain and its coverings.

Megrims or " vertigo," also spoken of as " staggers," occurring in harness or draught horses is almost always due to mechanical impediment to the flow of the blood from the brain, occasioned by the pressure of too tightly or badly fitting harness. By some, megrims is believed to be due to inflammatory action, but there do not appear to be any grounds for this supposition. It is said that megrims may be produced by exposure to the rays of the sun, or by driving fast after a heavy meal. These causes certainly may increase the tendency to this affection, but it is very improbable that they alone can cause it. As already pointed out, indigestion is liable to be caused by driving fast after heavy meals, and may induce dizziness or staggers, which it is not easy to distinguish from megrims depending on actual congestion of the brain.

An attack of megrims is generally sudden in its onset, there being usually no warning symptoms. The animal slackens speed, or stops

---

*Our readers will understand that the disease termed sprained back has no relation to true sprain of the muscles of the back, of which we shall treat along with other sprains.

suddenly, and moves the head from side to side, or up and down. Sometimes the horse turns its head to one side. The vessels of the face and throat are engorged, the eyes stare, the nostrils are widely opened, and the breathing is rapid. The skin may be bedewed with perspiration, and the muscles of the face twitch convulsively. If the collar causing the obstruction be removed, the symptoms abate, and the animal soon recovers. When the symptoms are very severe, there is great excitement, the convulsions become still more marked, and the animal falls prostrate to the ground.

It is necessary first to remove the collar, to permit of the return of the blood to the heart, and then to apply cold water to the head. When we have reason to suspect that the affection depends upon indigestion, this must be treated as we have already directed. When the neck is peculiarly shaped, it may be necessary to use a breast strap, instead of a collar. Mad staggers is nearly always due to inflammation of the brain, though frenzy or uncontrollable fury may be one of the symptoms of rabies, and sometimes has been thought to come on as a result of acute indigestion.

MAD STAGGERS, OR INFLAMMATION OF THE BRAIN, OR ENCEPHALITIS, OR PHRENITIS.

INFLAMMATION of the brain is a rare disease in the horse, and is usually due to direct injury, such as a blow on the head, but may also be caused by great exhaustion or exposure to the rays of the sun. Sometimes the symptoms are very sudden, consisting in great excitement with convulsions, followed by a stage of depression. At other times, the stage of excitement is absent. In these cases, the animal is very intolerant of its head being handled, or pressed upon, and the skin and mouth are hotter than natural. The eyes are staring, and the pupils contracted, though in the later stages of the disease they become widely dilated. The pulse is quickened, and the horse moves to and fro sullenly, and his body is sometimes bedewed with perspiration. Occasionally muscular twitchings and general or local insensibility are manifested.

The stage of excitement is of variable duration, and the symptoms manifested in it differ widely in intensity. It is followed by the stage of depression. Cases of inflammation of the brain call for all the care of the scientific veterinarian, and it is therefore impossible for the amateur to take such cases in hand. Bleeding is indicated when the fever is high, and the excitement very great. Generally from two to three quarts of blood may be removed. A full dose of aloes should be given in the first instance— say five to seven drachms, according to the size of the animal. Locally, ice or cloths steeped in cold water or some evaporating lotion (alcohol one part, solution of subacetate of lead one part, water eight parts), should be applied to the head during the stage of excitement. The animal should be removed from all noises, and kept as strictly quiet as possible. The diet should be light and nutritious. If the animal continues to drink, two drachms of bromide of potassium and two of hydrate of chloral, may be given every four hours in the water, during the stage of excitement. If paralysis continues after the abatement of the acute symptoms, a smart blister may be applied to the poll, and repeated if necessary.

EPILEPSY.

EPILEPSY is a rare disease in the horse. It may be defined as an affection of the nervous system, characterised by sudden temporary loss of consciousness, associated for the most part with a convulsive attack, which in many instances cannot be referred to actual disease of the brain. A horse when attacked with epilepsy, champs his jaws, becomes unconscious, and falls to the ground convulsed. Sometimes the spasms are very slight, and the animal quickly regains consciousness, and seems as well as ever. Sometimes the spasms are confined to one limb, sometimes to one side of the body, or to the muscles of a particular part, as the face or neck. The animal froths at the mouth, grates the teeth, moves the head quickly to and fro, and turns about wildly. During the attack, cold water may be dashed on the head, and all means should be adopted to prevent the horse harming himself in his convulsions. In very strong animals, bleeding has been practised. After the attack is over, the general health should be promoted, the diet carefully regulated, and the bowels opened. If the disease depends upon worms, these should be expelled. In chronic cases one drachm of each of the bromides of sodium, ammonium, and potassium may be given three times daily in the drinking water, for a week or two.

PARALYSIS OF THE LIPS.

THE only form of paralysis of which we need treat here, is paralysis of the lips, a disease not uncommon in horses. The nerves which supply the muscles of the lips are liable to become pressed upon by badly fitting bridles. Sometimes the nerve of one side, sometimes those on both sides become thus pressed upon, and paralysis ensues. When both nerves are affected, the lips cannot be closed, but hang pendulously, and saliva flows from the mouth. When the nerve of one side only is implicated, the lip, having no longer any power, is drawn by the action of the opposing muscles towards the other side. The horse cannot grasp his fodder when the lips are paralysed, and so he has to snatch his food with his teeth. In such cases the first thing necessary is to remove the badly fitting bridle, and to apply a blister of equal parts of the ointments of red iodide of mercury and of cantharides below the ear and along the cheeks. Internally, a moderate dose of aloes may be administered, and an eight drachm ball, made of two drachms of iodide of potassium, one drachm of powdered nux vomica, made up with a sufficiency of ginger and treacle, may be given twice daily. The diet should be soft and laxative, consisting of oatmeal and linseed cake gruel. If desired, the iodide of potassium may be given in the food, instead of administering the balls. Two drachms may be thus given twice daily.

HYDROCEPHALUS, OR WATER IN THE BRAIN. TUMOURS IN THE BRAIN.

WATER in the brain or hydrocephalus is not uncommon as a congenital defect in foals, but is only rarely met with in older animals. The hydrocephalic head is recognised by the great enlargement of the volume of the skull. In the early stages, the foal is irritable and feverish. Afterwards he becomes weak, and the sensibility is impaired. Paralysis and convulsions precede death in fatal cases. The largest amount of fluid recorded as having accumulated in a foal's brain, is two and a half gallons. Recovery in this disease is very rare, and even in the most favourable instances, there is little profit to be derived from keeping hydrocephalic foals, as they never thrive.

Tumours in the cavities of the brain of the horse are very common, but as they grow very slowly, and do not occasion severe symptoms until they have attained a size about as large as a pigeon's egg, their presence is rarely suspected until shortly before leading to a fatal result. At the autopsy of the famous racer Macgregor, Mr. Charles Gresswell, of Nottingham, found a large tumour in each of the lateral cavities of the brain.

CHAPTER VII.

POISONING.

Arsenic. Aconite. Ergot (Claviceps Purpurea); Grass Staggers. Lead. Hellebore (Veratrum Album). Antimony; Opium. Savin. Bryony. Cantharides or Spanish Fly. Euphorbium or Spurge. Yew Tree. Water Drop Wort. Meadow Saffron (Colchicum Autumnale). Remarks on the Condition of Horses.

ALTHOUGH it may be fairly stated that poisoning in horses is not so frequently met with as it once was, it is still common, and is, therefore, of very great practical importance.

In almost all cases of poisoning, it is noteworthy that the drug has been administered by the attendant with the intention of preventing or curing some real or imaginary disease, which the horse is supposed to be suffering from, or of promoting his well-being by increasing his appetite, or in other ways ; and it may be pointed out that whereas formerly mineral agents, such as arsenic and antimony, were largely given for these purposes, we now find that vegetable poisons, such as hellebore and overdoses of aconite, are frequently substituted. It is well known that many vegetable poisons are quite as powerful as the mineral ones, and we should, therefore, be especially suspicious of nostrums advertised to contain no mineral poison, for these but too frequently contain vegetable poisons still more dangerous. A large number of old formulæ in the hands of those employed in the stable, and on the farm, contain overdoses of arsenic, hellebore, aconite, antimony, and other preparations, which are seldom employed by the veterinarian except in severe cases, and some of them are scarcely ever given by him internally. Sometimes, however, more especially in the case of lead, poison is taken accidentally. At other times, though very rarely, it is given with criminal intent.

We will first consider the baneful effects produced by acute and chronic arsenical poisoning, and will then treat of the others in the order of their importance, at the same time mentioning shortly the treatment to be adopted in these cases.

ARSENICAL POISONING.

ARSENIC is usually administered to horses in the form of arsenious or common white arsenic anhydride. Though poisoning by this substance is of less frequent occurrence than it once was, arsenic is still very commonly given by labourers and waggoners, and more rarely by grooms, in certain parts of the country. When given in excessive doses, it is generally through ignorance that this is done; but instances are recorded of cases in which it has been given with criminal intent. It is usually made up in the form of a ball with soap, tar, or sulphur, or indeed any suitable substance. Sometimes it is administered as a powder in the food or water, and though the proper medicinal dose is but four grains, attendants commonly give as much as will lie upon a sixpenny or shilling piece, or even more. The following accounts will serve to show some of the more important symptoms and *post mortem* appearances of arsenical poisoning :—

When summoned one morning, some time ago, at 3 a.m., the late Mr. D. Gresswell found four cart horses in a very dangerous state. They were fine heavy animals in excellent condition, and on the previous day had shown no signs whatever of ill health. Their restlessness had attracted attention about 12 or 1 a.m. They were breathing rapidly, and the pulse was very rapid and almost imperceptible, the arteries. feeling like mere threads. All four animals were in great pain. They got up and down alternately, rolled over and over, and manifested other signs of intense agony. The bowels were very loose, and there was much straining. The extremities were cold, and the eyes were staring; and there was total loss of appetite, and extreme prostration. Eructations of gas frequently passed from the stomach. One horse died at 10 a.m., a second at 4 p.m., and a third at 10-30 p.m. Before death the animals became still more restless, the pulse was weaker and finally imperceptible; the mouth became clammy and the breath fetid, and they succumbed at length in a state of extreme agony and collapse. One animal recovered, but remained so weak and debilitated as to be incapable of rising without assistance. At length, however, he made a gradual and apparently complete recovery, but was not able to resume work for three or four months. When the stomachs of the animals which had died were examined, they were found to contain undigested food, and the contents were tinged with blood. The membrane lining the stomach was blackened, and in parts the walls were much eroded, forming many large eschars or patches of burnt tissue, and in other places the lining was raised in the form of small blisters. In one of the cases there were two almost complete perforations through the walls.

In these cases, although the waggoner denied having administered anything, it was afterwards elicited, that he had given to each of the horses a quantity of the white arsenic, made into balls by mixing it with tar. This he had given at about 8 or 9 p.m. the previous day.

On the 20th of June, 1883, we had a team of four cart horses belonging to a farmer, under our care. The symptoms in these cases were similar, but much less severe, than those above described. One of the animals died, but the remaining three made a gradual recovery, and were soon again at work. It was ascertained in these instances that the drug had been given in the form of the ordinary white arsenic. The waggoner had for some time previously given to each of his horses every night, as much as he could place on the end of a large pocket knife. On the night when the horses were so suddenly affected, he had given an extra dose to each, three or four hours previous to the appearance of the symptoms of poisoning. Several months afterwards, we took the opportunity of examining two of the horses which had recovered. Both were found to have diseased hearts, and the foreman informed us that they never regained their previous strength.

That arsenic when given in solution acts much more rapidly and powerfully, is shown by the following record of nine cases of poisoning, which occurred in the late Mr. D. Gresswell's practice some years ago. One of the waggoners on a large farm having obtained a pound and a half of white arsenic, stirred it in a tub of boiled linseed gruel. This was served out equally to nine horses, on their return from work, at two o'clock in the afternoon. Very shortly afterwards the horses manifested considerable uneasiness, and eight of them died very quickly, while the ninth recovered under very careful treatment and management.

We might record many other cases, but the above will suffice to illustrate the baneful effects of arsenic. Before closing our remarks on the subject, we must say a few words concerning chronic arsenical poisoning. At the present time, this form of poisoning is of much more frequent occurrence than the acute form; and although sometimes the horse may escape any outward signs of indisposition from the occasional administration of small overdoses of white arsenic, yet the practice of administering this drug by attendants, is to be deprecated from every point of view, as it not unfrequently totally incapacitates the animal from any prolonged exertion. In February, last year, we were called to see a valuable seven-year-old hunter, belonging to a gentleman residing on the Lincolnshire wolds. The horse had an excellent appetite, but was in poor condition. The pulse was fairly strong, but irregular, losing a beat every now and again. The breathing was somewhat accelerated. We were informed that when galloped even for a short distance the horse breathed laboriously, and could only with difficulty be induced to go beyond a slow trot. It was ascertained that for many months previously, the late groom had given to the horse small doses of arsenic at regular intervals. The untoward symptoms were attributable to this practice, as the horse had always enjoyed perfect health previously, and made much improvement after the groom left.

Arsenic should not be given unless for some definite object, and, when necessary, is best administered in the form of Fowler's solution, of which the dose is half an ounce in the drinking water after meals. Arsenic has a special action on the skin, and is very useful in many forms of skin diseases in horses and other animals. It is mainly given by attendants to make the coat more glossy and smooth, and it is a common ingredient in the alterative

balls prepared from recipes in the possession of many stablemen and waggoners.
We believe that when given it is with much more caution than formerly;
but this cannot be said of some of the poisons. In almost all cases where
arsenic is given as an alterative by attendants, half an ounce of bicarbonate
of potassium, given once or twice daily in the drinking water, would be
equally efficacious, and without any danger. If an appetiser is wanted, a
ball may be made of equal parts of carbonate of ammonium, ginger, and
gentian, made up to one ounce with treacle. This is found very efficient.

ACONITE POISONING.

ACONITE, which is one of the most active and valuable of the pharmacopœial
remedies employed, is a common cause of poisoning in the horse, and
is certainly on the increase.

It is not generally known that many quack nostrums and some formulæ
in the possession of stablemen and others contain overdoses of tincture of
aconite. When the doses are administered in rapid succession, very alarming
symptoms are produced. Frequently cure of the animal is rendered well
nigh impossible. We have often been sent for to horses, in cases where
sudden difficulty of breathing and gurgling in the throat have supervened
from the administration of aconite. These symptoms generally subside
quickly—when the overdose has not been excessive—on the administration
of spirit of ammonia and brandy. The drenches which contain aconite
in the form of tincture, are generally those called inflammation drinks. It
must be remembered that the dose of Fleming's tincture of aconite is only
from five to ten drops, and of the ordinary tincture of aconite thirty to
forty drops. Such doses should not be repeated more frequently than
once every three or four hours. During the past two years, we have had
more cases of poisoning by tincture of aconite than by any other poison.
The owners in these cases often seem not a little surprised when informed
that their animals are suffering from aconite poisoning.

Only a short time ago a valuable horse was poisoned by the groom,
who kept tincture of aconite by him for use at his own discretion. Two or
three drachms were not thought too much to give, and although death
followed in about an hour, and the animal gasped for breath at the feet of
this attendant, the fatal event was attributed by him to occult influences
of an inflammatory kind.

The special symptoms manifested in horses which have received an
overdose of this active drug are the following :—The breathing becomes slow,
feeble, and more difficult, the animal trembles all over, and there are not
uncommonly gurgling sounds in the throat, and frothing at the mouth,
sometimes succeeded by convulsions. Perspiration bedews the surface of
the body, and the pulse becomes weak, and sometimes almost imperceptible.
In some cases we have known the animal fall to the ground from absolute
loss of power to stand, and in rare instances he manifests great restlessness
and pain.

A good formula for ordinary inflammation drenches, which is at once safe and efficient for those purposes for which these draughts are commonly employed, is the following :--Of liquor ammonii acetatis four ounces, of Fleming's tincture of aconite five drops, of spirit of nitrous ether one ounce, and water added to make half a pint.

POISONING BY ERGOT.

THE next poison—ergot—of which we shall treat is one of some importance and interest, not only as affecting the equine tribe, but also as a source of disease among cattle. Ergot, or ergot of rye, is caused by the growth of a fungoid parasite which infests a number of grasses and cereals, more especially rye. The cultivated grasses which most generally become diseased by the growth of the vegetable fungus called *Claviceps purpurea*, are timothy grass, tall fescue, floating sweet grass, fox tail, and rye grass. The weed grasses most generally infested with the parasite are soft brome grass, meadow brome, couch grass, and wall barley grass. The ergot itself is a purplish or bluish black, hard, elongated body, easily recognised again when once carefully observed. In those parts where rye-bread is much eaten, ergot is often present in large quantities in the flour, and very alarming symptoms, and sometimes even death, results in those who have partaken of it. In Russia, gangrene, or mortification of the limbs and other parts, has, especially in certain seasons, resulted from this cause. Ergot is not uncommonly a cause of abortion in mares and cows; and it is recorded that a Shropshire breeder of cattle lost £1,200 in three years, from the grasses in his pastures becoming ergotised.

The disease termed grass-staggers, produced in horses by feeding on rye grass at a particular period of its growth, appears as a local affection, when horses are grazed on land where this abounds. This affection appears to have some resemblance to ergotism. The symptoms are gradually developed, and the animal manifests deficient controlling power over his muscles, especially those of the hind extremities. The weakness gradually increases, and the horse reels or staggers. Muscular spasms are occasionally manifested, and when the animal falls, they are sometimes very severe. Consciousness becomes impaired, and death sometimes terminates the malady. This disease must not be confounded with stomach-staggers or acute indigestion, of which we have already spoken. Grass-staggers is rarely fatal when the cases are attended to in the early stages. The animal should be removed to a fresh pasture, as soon as the disease shows itself, when recovery will in most cases follow without further treatment.

LEAD POISONING.

LEAD poisoning is generally confined to certain districts where lead smelting is carried on, but it may also occur in horses from ingestion of lead paint, or splinters of bullets, which are scattered about near rifle targets. In the pure metallic form lead appears to be devoid of poisonous properties, and it is well known that in the form of shot it is used by dealers of questionable principles to alleviate the symptoms of broken wind in horses they have for sale.

In some instances, lead poisoning has been due to boiling food in vessels used for containing lead preparations. It is known also that lead may be absorbed by water conducted through pipes of this metal, and this is more especially likely to be the case when the water is highly oxygenated, or contains organic matter or certain gases. Lead poisoning in horses may be acute, when it is spoken of as saturnine epilepsy, a disease in which stupor, delirium. or convulsions are manifested, or it may be chronic.

HELLEBORE POISONING.

HELLEBORE poisoning was some years ago very common, this drug forming one of the most common ingredients of the powders and balls of stablemen and quacks. Even now, cases of poisoning by this dangerous drug are not rare. It is supposed by grooms to have a valuable alterative effect, but the idea is a mistake. In a case recently under our notice, two drachms of the powdered hellebore root were given by the groom to a carriage horse. When called in to see this animal, the writer found the head protruded. The pulse was much accelerated, and varied from 90 to 100 beats per minute, the respirations were much quickened, the extremities were deathly cold, and there were marked nausea, and frequent attempts at vomiting. The appetite was completely lost for forty-eight hours, after which it gradually returned, and the animal made a slow but complete recovery.

The late Mr. D. Gresswell saw a large number of cases of hellebore poisoning at different times, and in some, actual vomiting took place. This occurrence, as is well known, is rare in the horse. Hellebore poisoning is frequently mistaken for choking by the uninitiated ; but the history of the case—when that is to be obtained—and the character of the pulse will at once distinguish it from this accident.

On March 6th, 1886, we were asked to see a heavy draught-horse said to be choking. The symptoms observed by the owner had supervened three hours after the administration of a ball containing a large quantity of hellebore (Veratrum album). It is almost needless to add that on our arrival, nothing whatever was told us concerning the ball which had been given for the purpose of curing the grease, from which the animal was suffering. This information was elicited by close cross-questioning. In this way it was discovered that the balls had been procured from a chemist. The animal was retching continually, but there was no actual vomition. The pulse was very irregular and feeble, and numbered eighty-six beats in the minute. The respirations were sixty-eight. The symptoms had gradually been becoming more severe, until when death seemed imminent, help was sought. Three ounces of whiskey, together with three ounces of solution of carbonate of ammonium. were ordered to be given every hour for six times, and then every two hours. In twelve hours the animal began to improve. On the following day he was much better, and tonics were thereupon substituted for the stimulants. The horse rapidly recovered, and was soon well again.

POISONING BY PREPARATIONS OF ANTIMONY.

ANTIMONY is still not uncommonly administered to the horse in the forms of tartar emetic and butter or chloride of antimony, which often constitute main ingredients of the recipes for balls and powders in the possession of stablemen and grooms. Antimony preparations are not nearly so frequently given as they once were, and the practice is no doubt becoming still more rare. The late Mr. D. Gresswell had a large number of cases of poisoning by these agents under his care in the course of his lifetime, and the writer has had a few examples, which fortunately, however, did not prove fatal.

When a horse has had a large dose, there are manifested frequent attempts at vomiting, and this may actually occur. The pulse becomes weak, fluttering, and almost imperceptible. There is great prostration, and gradual loss of consciousness in severe cases, followed by death.

POISONING BY OPIUM.

OPIUM is not a common cause of poisoning in the horse, but it is sometimes given by dealers and others in poisonous doses, in order to prevent kicking and restiveness in horses they wish to sell. A few months ago, the writer attended a half-bred mare to which the owner had administered one ounce and a half of Turkey opium. When called in on the day following the administration of the drug, the mare was found to be in a very dull, dejected condition, and the pulse was very feeble and soft. The pupils of the eyes were contracted to pin points, and the membrane lining the nostrils was of a darkish brown hue. The symptoms remained unabated for three days, during which time the animal continually moved round and round in the box. On the fourth day, the pulse began to regain vigour, but recovery was not complete before the lapse of a week.

MERCURIAL POISONING.

THE next agent of which we shall speak is a very poisonous preparation of mercury called corrosive sublimate. It is sometimes given by stablemen in injurious doses to horses, causing loss of appetite, salivation, pawing, looking at the flanks, rolling about, profuse perspiration, rapid and weak pulse, violent action of the bowels, straining, convulsions, and death. On no account whatever should this excessively dangerous drug be used.

POISONING BY SAVIN, BRYONY, CANTHARIDES, EUPHORBIUM OR SPURGE, YEW TREE, AND MEADOW SAFFRON.

SAVIN is another drug sometimes given by grooms and others with the idea of improving the general condition, and death has often been caused by this practice. It is said that the presence of savin can be detected in the stomach of the dead animal, by the black-currant-leaf like smell of the contents when boiled in a little water and beaten up in a mortar.

Bryony also is often given by horse-breakers to young animals with a similar idea, but, although this drug excites the poor creature, and for a time appears to improve his condition, it is, nevertheless, decidedly poisonous, and when the transient effects are over, depression and loss of condition follow.

Cantharides or Spanish flies are sometimes administered by attendants, and owing to the large amount sometimes given, death has sometimes resulted. Its use by amateurs is in every way to be deprecated.

Euphorbium or spurge, one of the components of the old farriers' blister, has also caused many deaths, which have resulted from the great irritation set up by this drug.

Of the remaining poisons, yew tree, water drop-wort, and meadow saffron, which are sometimes eaten by horses out at grass, our readers probably have some knowledge.

Many instances of death from browsing on the leaves of the yew tree (Taxus Baccata) have been recorded. After death, which in some cases takes place in from two to three hours after the ingestion of the foliage, the stomach has been found contracted and inflamed. The method of treatment to be adopted in cases of yew-tree poisoning, is the administration of a pint of linseed oil, with two ounces of spirit of ammonia, and one ounce of nitric ether. In a couple of hours this draught may be repeated, and again after an interval of four hours, the ammonia and ether may be given alone in a pint of gruel.

The water drop-wort is a plant which grows in ditches and marshy localities. This plant is not often eaten by horses, but brood mares with vitiated appetites have been poisoned by ingesting it.

The meadow-saffron or autumn crocus, known botanically under the name of Colchicum autumnale, is sometimes a cause of death to horses and cattle. Several cases of poisoning by eating the stalks, leaves, pods, and seeds of the plants, have been recorded, but the writers have never had under treatment a case of poisoning by this vegetable. The symptoms manifested generally, are colic, and great dulness, followed by death in about twenty-four hours. At the autopsy, the stomachs have been found inflamed and eroded. Cattle, when poisoned by this plant, present pretty much the same manifestations, viz. :—colic, diarrhœa, great straining, dulness, cold extremities, and extreme prostration. In these cases, it is best to give mucilage of linseed, with one ounce of spirit of ammonia, and three ounces of brandy, repeated every two or three hours.

Here ends the list of poisons.

ON REMEDIES WHICH CAN BE SAFELY GIVEN IN ORDER TO PRESERVE CONDITION.

WE may now show how the preceeding poisons administered with the view of improving the condition and acting as alteratives may be dispensed with, and their place taken by remedies at once more efficient and not dangerous. We must remember, first and chiefly, that medicines cannot alone bring about that healthy condition which it is our object to secure, but they can be of great service in aiding other measures adopted to attain this end. Moreover, it must always be borne in mind, that when the condition is satisfactory, proper measures should be taken to secure the maintenance of health. Enforced idleness, over-work, over-feeding, under-feeding, insufficient air, over-crowding, disease, and pain, are all antagonistic to the preservation of condition. The cause of the loss of appetite should be ascertained, as sometimes this may proceed from irregularities of the teeth, which may require rasping, or other treatment.

If the skin is out of order, and grease or humour manifest themselves, one may administer a full dose of aloes, and afterwards balls made of one drachm of grey powder and gentian to eight drachms, given twice daily. In addition, half an ounce of bicarbonate of potassium may be given twice daily in the drinking water. It must always be remembered that after a full dose of aloes, a horse requires three full days' rest and bran mash diet.

Very commonly, from some cause or other, the attendant perceives that the animal would be all the better for some alterative medicinal treatment. It is hardly necessary to say that the cause of "indifferent condition" should first be inquired into, in order that if possible it may be rectified. Is the food good? Is it in proper amount and of good quality? Are the hay and oats good? Are the meals given regularly? Are the bowels too costive, or the reverse? All these questions present themselves for consideration.

As a general alterative, a table-spoonful of powder, composed of four parts of precipitated sulphur, four parts of nitrate of potassium, one part of gentian, one part of fenugreek, half a part of carbonate of iron, with a little essential oil, such as oil of cajuput, one-sixteenth of a part, may be given once daily in the food. Or we may give eight drachm balls, composed of resin five parts, nitrate of potassium one part, gentian two parts, carbonate of iron one part, cubebs one quarter of a part, aniseed one quarter of a part, made up with oil of turpentine, and soft soap. These balls may be given every other day, or every day at first, for a week or so.

Before commencing with condition powders or balls it is well to give a moderate dose of aloes, from three to six drachms. The aloes should be of the best quality. The practice of administering small doses of aloes in every alterative and condition ball is to be strongly deprecated. A far more wholesome practice, and one of great advantage, is to give horses a good bran mash, twice weekly in the evening, after the day's work is over.

If the appetite is bad, and a general stimulant and appetiser is needed, equal parts of carbonate of ammonium, ginger, and gentian, made up into an eight drachm ball with treacle, may be given at first twice a day for a week or two, and then once a day for a week.

Part II.

Surgical Disorders of the Horse.

CHAPTER I.

DISEASES OF THE SKIN.

General remarks on the functions of the Skin. Mange. Dermatodectes Equi. Sarcoptes Equi. Symbiotes Equi. Ringworm. Urticaria, or Surfeit. Hide-bound. Eczema. Cracked Heels. Grease and Grapes. Mallenders and Sallenders. Mud Fever. Warts.

GENERAL REMARKS ON THE FUNCTIONS OF THE SKIN.

As the diseases of the skin of the horse are very numerous and varied, it is our purpose to enter pretty fully into the consideration of their symptoms and treatment. Before, however, commencing our description, we may say a few words regarding the functions and structural peculiarities of this important covering, for these are of interest and are well worthy of a few moments careful attention.

The skin is described as a soft and pliant membrane, which invests the whole of the external surface of the body, following its prominences, its depressions, and its curves. It serves as an effectual protecting cover, preventing the penetration of noxious materials, and allowing of the escape of effete matter in a gaseous liquid and solid state from the blood. The skin also has other important offices, for it acts as a sensitive organ in the exercise of touch ; while it plays a very important part in keeping the temperature of the body constant. This varies in health in the horse from 100° to 101° F. Our readers are aware how the blood vessels of the skin become contracted in cold weather, and how, on the contrary, in hot weather they dilate, when perspiration is excreted in much larger quantity. By the evaporation of the sweat passed out, heat is absorbed, and thus the bodily temperature does not increase materially on the hottest summer's day. Exercise increases the production of heat in the body, but it also increases the rapidity of the circulation in the blood vessels, which become dilated, and thus the sweat glands of the skin become more active. By the perspiration excreted and vaporised on the surface of the body, heat is prevented from increasing above the standard in health. The actual quantity of water excreted per day by the skin in the shape of watery vapour is very large. In the human body it varies from a pint and a half, to two pints. Finally, also, the skin acts as an absorbing organ.

1

The skin consists principally of a layer of vascular tissue called the derma, and an external covering called the cuticle. Within and below the derma are embedded the sweat glands, which excrete the perspiration, the sebaceous glands which secrete the oily fluid to lubricate the skin, and the little depressions called hair follicles, in which the hairs are situated. The hair and nails, strange as it may seem, are merely modifications of the cuticle or epidermis. The upper surface of the derma is not level, but shows a multitude of little elevations which are termed papillæ, in which the little nerve endings terminate, thus endowing the skin with sensibility. On the tips of the fingers of man and in other parts, which are endowed with extra sensibility, the nerve fibres ending in the papillæ are more numerous than elsewhere. Likewise, on the tip of the nose of the horse and in other parts, they are more abundant. The papillæ are about 1·200th of an inch in length, and about 1·600th of an inch in width at the base. The cuticle is a thin layer covering the derma, and filling up the depressions between the papillæ. It is made up of little cells, which are being continually deposited on the derma.

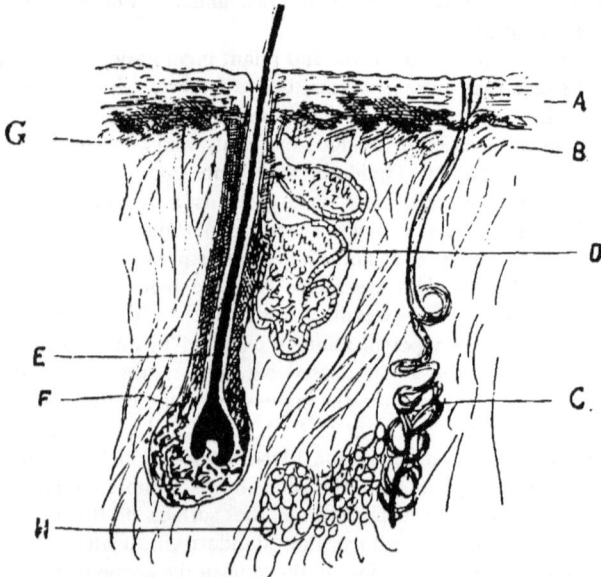

In the above picture of a section of the skin of the horse A is the cuticle, B is the derma, C is a sweat gland, D is a sebaceous gland, E is the hair in its follicle, F is the hair bulb, G is a papilla, and H is a group of fat cells. The section is magnified highly.

In the horse, the bristly appendages known as horse-hair, should be distinguished from the other hairs forming the coat. The latter are fine and short, especially in the regions where the skin is thin, and where the hairs are imbricated on each other. The former are thicker and longer, those of the tail being the longest and strongest on the body. Those which form the "foot locks" are peculiar to the horse, and vary in length and coarseness with the breed of the animal. When hair is fine and long and wavy, it forms wool; and when straight and rigid, as in the pig, it is known as bristles (Chauveau).

The sweat glands are very numerous over the surface of the body. They consist of small lobular masses formed of a coil of a gland tube surrounded by little blood vessels, and embedded in fatty tissue in or beneath the derma. From the coil passes a duct, which opens on the surface of the skin. According to Erasmus Wilson, there are as many as 3,528 glands on each square inch of the palm of the hand of a man, while on the neck and back they only amount to 417. The total number of these glands in the human being is estimated at nearly two and a half millions.

MANGE.

AFTER these few preliminary remarks, we may at once proceed with the consideration of the symptoms and treatment of the various maladies of the skin of the horse. We shall first devote our attention to the parasitic diseases. These fall into two main groups, viz., those due to animal parasites, and those due to vegetable parasites belonging to the order of the fungi, such as the various kinds of ringworms.

Mange or scab is an affection of the skin, decidedly contagious, caused by the presence of little creatures belonging to the same order as the mites. These little animals are of three varieties in the horse, but do not differ very much in appearance or size. The "scab" acari of slightly differing kinds infest man and all the domesticated animals. These parasites are said to live on the fluid, which is effused from the blood, owing to the irritation their presence sets up.

The first kind which infests the horse, termed dermatodectes equi, of which we append a drawing, is the kind most frequently met with in England. This creature causes the formation of little elevations on the skin, in the upper part of which the contents soon become liquid and burst, and afterwards becoming drier, form crusts or scabs. These little elevations or pimples, which are about an $\frac{1}{8}$th of an inch in height, are especially numerous on the upper part of the neck and root of the tail. If a few crusts be taken off, placed on a white surface, and exposed to the heat of the sun, the parasites may easily be discerned with a small magnifying glass. The itching which is set up by these creatures is of an intense character. They deposit a secretion of great acridity, and by their long mandibles or jaws, they cause serious alterations in the skin, which is rendered bare, wrinkled, and bleeding, especially around the mane. The disease occasioned by them is more amenable to treatment, and spreads much more slowly than that produced by the second variety. The greater facility with which these creatures can be killed by the application of ointments, is no doubt due to the fact that they do not burrow into the skin, but merely conceal themselves under and among the scabs. The dermatodectes live in colonies.

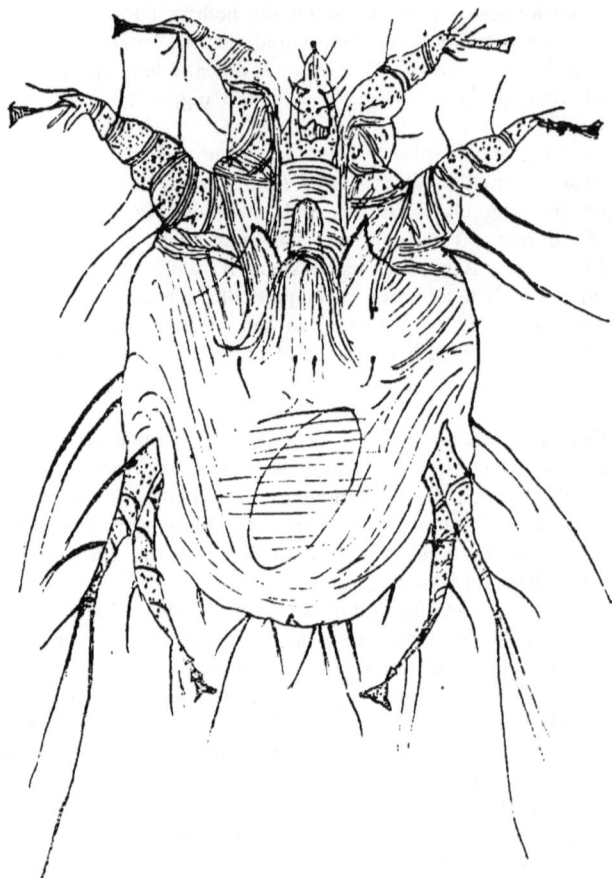

Dermatodectes equi (Gerlach).

The second variety of mange is caused by the sarcoptes equi These creatures penetrate the skin, raising up a small knotted elevation, with a small passage, at the extremity of which the acarus resides. This acarus has a tendency to wander about, and is especially abundant on the sides of the neck and withers, from whence it spreads over the surface of the body, excepting those parts covered by long hair. The pimples, if examined, are found to be hard scabs, situated on a moist basis. As the crusts become drier, the skin becomes thickened, wrinkled, and fissured. Mange caused by this acarus is rare. The course of the disease is slow, and may even cause death from irritation and exhaustion. Dr. Fleming, F.R.C.V.S., LL.D., has seen the disease in the Crimea. The sarcoptes do not live in colonies like the foregoing, but lead an independent existence.

The third variety of mange is caused by the symbiotes equi, a creature which lives in colonies and invades the limbs, not burrowing, but merely biting through the skin, and leading to the exudation of fluid, which forms large scabs. This variety of mange is not so contagious as the previous ones.

We may now consider the treatment of the various kinds of mange. In the first place, the horses affected should be isolated from the healthy ones. After cleansing thoroughly with soft soap and warm water, the affected parts may be smeared over with sulphur ointment twice daily. A still more efficient ointment for the cure of mange we may append from the "Veterinary Pharmacology and Therapeutics." It is made of one ounce of ointment of sulphur, one ounce of ointment of stavesacre, one drachm of white precipitate of mercury, and twenty drops of carbolic acid or creosote. This ointment we have found very efficient. Professor Williams, in his excellent work on veterinary surgery, recommends the following ointment as most effective: --Of powdered stavesacre two ounces, of lard eight ounces, of olive oil one ounce. Mix and digest at 100° in a sand bath, and strain.

In addition to dressing the diseased parts of the skin, it will be necessary to cleanse very carefully the clothing and fittings of the affected animal. The rugs may be steeped in boiling water, to which has been added soft soap and carbolic acid. The fittings should also be thoroughly washed and cleansed with warm water and a solution of carbolic acid. Williams recommends that the harness, saddle, and grooming utensils should be washed with soap and warm water, and afterwards with a solution made of ten grains of corrosive sublimate to each ounce of water. This substance, however, is very poisonous, and if used must be employed with great caution.

The horse, in addition to being attacked by scab, is also liable to be attacked by lice or pediculi. These insects occasion very violent itching, which increases at night. This disease is termed poultry lousiness, because it is from ill-kept poultry that the insects gain access to the stable. This disease—which is easily cured when the cause is remedied by removing the poultry and cleaning the stable—is characterised by the eruption of a number of small blebs on the skin. These cause the hair to fall off in little round patches, about the size of a pea or bean. In these cases all that is necessary is to remove the cause, cleanse the hen-houses, whitewash the stable, and wash the animal with a solution made by boiling one ounce of stavesacre seeds in a quart or so of water.

RINGWORM.

RINGWORM is a disease of the skin caused by the growth of vegetable parasites, belonging to the order of the fungi. These little plants, of lowly form and structure, are of two varieties, and give rise to two apparently somewhat similar, but nevertheless really different, forms of ringworm. As, however, the treatment of the two diseases is in the main similar, the diagnosis is not a matter of great moment.

· The first variety of parasite causes the ordinary or common ringworm known as tinea tonsurans, a very common disease in man and in the domesticated and other animals. Unlike the other form of ringworm termed favus, a very much rarer malady, it is not especially liable to attack debilitated animals. We shall first devote our attention to the consideration of the common variety, and then shall shortly review the nature and treatment of the much rarer form.

In the ordinary ringworm, the hairs are invaded in circular patches by the rapidly spreading fungoid growth. If this parasite be examined under the microscope, it will be seen to consist of little slender-jointed rods, and small highly refractile spores. This parasite spreads not only into the sheath, but also up the shaft of the hairs, and is known technically as the trichophyton tonsurans. The hairs become drier and more friable, and then break off near the roots, leaving little bald patches covered as it were with stubble. The commoner seats for the growth of ringworm are the back, neck, hind-quarters, and face.

Ringworm is a very contagious malady, and the animals infected should therefore be isolated for a time from the healthy ones. To prevent the spread of the disease, the stables should be cleaned, and the walls whitewashed. The affected parts should be thoroughly washed with soft soap and hot water, and the scabs removed. After these preliminary steps, the circular patches may be dressed three times daily with a concentrated solution of hyposulphite of sodium (two drachms to each ounce of water); or, if preferred, the parts may be anointed with iodine ointment, or with a solution of blue vitriol (one drachm to each ounce of water). The harness, collars, and clothing should be washed with soft soap and hot water, and then with water to which hyposulphite of sodium has been added, in the proportion of one ounce to a quart of water. The rarer form of ringworm, generally spoken of as honey-comb ringworm, is due to the growth of a fungus called the Achorion Schönleinii. This disease is attended by the formation of yellowish cup-shaped scabs of a circular form. Professor Williams records that some years ago he was called upon to attend a number of animals affected with this yellow honey-comb ringworm. This disease, which may be communicated from man to animals, in this respect resembling the common variety of ringworm, had attacked twenty horned cattle, three horses, some dogs, and several cats. The latter creatures had been in the habit of sitting on the backs of the horses and cows, and "doubtless the disease had been caught from mice by the cats, and then transmitted by them to the other animals about the place." The formation of the circular patches is attended by some itching. The hairs are generally invaded in this form of ringworm, as in the other variety. The scabs should be thoroughly washed in warm water and soft soap, and then anointed with the hyposulphite solution, or with the official ointment of iodine, or that of tar.

The ointments are probably best adapted for general use in cases of ringworm, because they do not become so quickly dried, and one application a day will prove sufficient, whereas the solutions, especially that of hyposulphite of sodium require to be painted on the affected part at least twice or three times during the course of each day.

A preparation composed of one part of oleate of copper, with four or five parts of lard, is proving very useful in cases of ringworm, and is well worthy of further trial. In the accompanying picture A shows the filaments and the little round spores of Favus ; B shows a hair invaded by the fungus of the common ringworm, the round spores of which are seen covering its surface ; C shows the spores of the ringworm in filaments, and also more highly magnified separately.

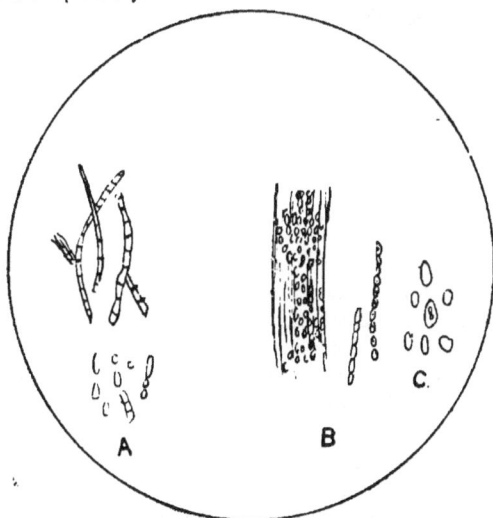

The only disease for which ringworm is likely to be mistaken is one called circumscribed herpes (herpes circinatus), which is an eruptive skin affection, characterised by the formation of rounded patches of little blebs. In ringworm, however, scales are found round the single hairs, or in patches surrounding several hairs. This is not the case with herpes, which is a non-contagious malady closely allied to eczema, and requiring the same treatment.

URTICARIA OR SURFEIT.

WE may now consider the various kinds of non-contagious skin diseases of the horse, viz.: surfeit, hide-bound, eczema, grease, cracked heels, mud fever, mallenders and sallenders, and warts.

By the term surfeit or urticaria, is understood a condition of the skin characterised by the eruption of a number of irregularly circular or ovoid elevations, or lumps.

These elevations are generally formed suddenly, and the parts most commonly affected are the loins, neck, and hind-quarters. Surfeit in almost all instances is due to impaired digestion, brought on by various causes. The affection is, as a rule, not characterised by great itching, though no doubt there is generally a certain amount of local irritation. Surfeit generally dies away in seven or eight days, and in most instances leaves no trace of its former presence ; though sometimes the hair which covered the elevations falls off. When the hair grows again, it is of a lighter colour on the spots from which it had fallen off. In these cases it is best to commence treatment by administering three or four drachms of aloes. If it is not possible to rest the animal for three days, a pint of linseed oil may be given instead ; for the administration of aloes always entails three days' complete rest. The diet should be laxative, and restricted in amount for two or three days. In addition to the physic, six drachms of bi-carbonate of potassium (and two drachms of Fowler's solution in bad cases) may be given twice daily for four days or so in the drinking water.

HIDE-BOUND.

HIDE-BOUND, though sometimes described as disease of the skin, is in reality merely symptomatic of a deranged condition of the system. Indigestion as in surfeit is one of the chief causes of this tightened condition of the integument, which, indeed, is common enough in many diseases of the horse. When we have reason to suspect that indigestion is the cause, the malady should be treated as described above ; and similarly if the animal has worms, these should be expelled. Where a stimulating medicine is required to brace up the system, eight drachm balls composed of carbonate of ammonium one drachm and a half, of citrate of iron and ammonium a drachm and a half, of powdered nux vomica half a drachm, of powdered capsicum ten grains, and made up with gentian and treacle, may be given every morning and evening for a week.

ECZEMA.

ECZEMA is an inflammatory disease of the skin characterised by the eruption of a number of small vesicles or blebs, the fluid contents of which escape, and congealing, form scabs. The cause of eczema is to be sought for in an altered condition of the blood, brought on by injudicious feeding. It may be due to interference with the normal action of the skin, owing to the wearing of uncleansed rugs, or to the accumulation of dirt which may irritate the skin, or to causes such as cold, which may check perspiration. Eczema may break out in almost any part of the body. The most common seats of this malady are perhaps the shoulders, the insides of the thighs, the neck and the sides.

In cases of eczema, it is best to commence treatment by the administration of three or four drachms of aloes, feeding the animal for three days on bran mashes and warm water, and resting him wholly during the time. The food should be laxative. Linseed cake, gruel, and hay or green food, with only a moderate amount of corn, may be substituted for the full allowance of oats. Internally, one ounce of bicarbonate of potassium may be given twice daily in the drinking water, and if the case be a severe one, two drachms of Fowler's solution may be given with it, in addition. The rugs, if woollen ones, should be changed and cotton rugs should be substituted, as woollen fabrics increase the irritation of the skin, and annoy the animal greatly. It is well to leave a piece of rock salt in the manger in cases of eczema, as it often has a very beneficial effect.

Finally, with regard to local applications for the affected parts, we believe the compound ointment of petroleum to be as good as any. The formula is of vaseline four ounces, of white precipitate of mercury four drachms, of liquor carbonis detergens four drachms. This ointment may be applied twice daily, and need not be rubbed off. The liniment of lead with oil is also a good application. It is made of half an ounce of solution of subacetate of lead mixed with four ounces of olive oil. When the itching is very troublesome, the parts may be dressed with a lotion made of four drachms of diluted prussic acid, two ounces of glycerine, and eight ounces of water.

CRACKED HEELS.

By cracked heels we understand a condition of the heels characterised by heat, tenderness, and little cracks, from which a serous fluid oozes. This affection is not usually attended by lameness, but when very pronounced, the animal not unfrequently is decidedly lame. Among the chief causes of this irritable, inflammatory, and painful state of the skin, which is more commonly encountered in thoroughbreds and hunters, are cold and dietetic errors. Cold or chill of this part of the skin, which is often only sparsely covered with hair, is sometimes caught while the animal is exposed for a long time in boisterous weather ; but is more commonly due to the practice of washing the legs after a day's work, and then not thoroughly drying them. Almost all the cases which have come under our notice are due to this avoidable source of error. It has been said that cracked heels are more likely to follow the use of hot than that of tepid or cold water, and this we believe to be true.

The practice of washing horses' legs with hot water after the day's work is over, is a favourite one among grooms, though it is rather frequently attended by evil effects. It may, however, be pointed out that if the parts were *thoroughly* dried after being washed, this would not be the case. We may also add, that a little simple ointment of vaseline, or of two parts of vaseline to one of glycerine, will prove beneficial in preventing this inflamed condition of the skin of the heels. Brushing and rubbing the legs is all that is necessary in order to clean them, and when the feet and legs are wet they should be thoroughly dried. If covered with mud, this should be allowed to dry on them, and brushed off next morning when dry.

Regarding errors in diet as a cause of cracked heels, we may mention that this condition not unfrequently follows the use of bad hay and mouldy oats. In some parts of the country, where the water contains a large amount of the salts of calcium or magnesium, cracked heels are more common than elsewhere. In slight cases of cracked heels, all that is necessary is the application of some simple soothing ointment once or twice daily. Ointment of boracic acid, is as good as any we are acquainted with. Take of bees' wax one part, paraffin two parts, almond oil two parts ; melt and add in fine powder boracic acid (warmed) one part ; mix and stir. Zinc ointment is also a fairly good application ; so also is an ointment of camphor one part, almond oil four parts, wax three parts.

When more astringent applications are required, ointments containing acetate of lead are valuable, such as almond oil six parts, and solution of subacetate of lead one part. Another valuable application, recommended in our "Veterinary Pharmacology, and Therapeutics," is the compound ointment of petroleum, made of white precipitate of mercury one drachm, liquor carbonis detergens one drachm, vaseline one ounce. Another good application may be made of citrine ointment four parts, almond oil two parts, paraffin two parts, and camphor one part.

Any of these ointments will relieve the irritable condition of the skin. They should be applied not only after the day's work is over, but also before the horse starts his day's work in the morning. It is rarely necessary to administer any medicine internally ; but, where the inflammation is very pronounced, it is well to give four or five drachms of aloes, and rest the animal for three days, in the meantime feeding him on mashes and warm water. Half an ounce of bicarbonate of potassium may also be given in the drinking water twice daily for several days. Locally it is best not to apply astringent ointments, so long as the part remains very red and inflamed, but to poultice it with bran for two or three days. In inveterate cases, when the part continues to discharge, we may paint it with a solution of nitrate of silver (fifteen grains to the ounce of water), once daily for two or three days.

GREASE AND GRAPES.

WE have now to consider the symptoms and treatment of grease, and of its more aggravated condition termed grapes. Not much is known of the actual pathology of grease, but of its causes and of the best means of curing it our knowledge is much more definite. Grease is an inflammatory condition of the skin of the limbs, characterised by heat, pain, and sometimes by lameness, and manifested by a sore or ulcerated condition, not uncommonly attended by manifestations of constitutional febrile disturbance. From the skin there oozes a thick, serous, oily discharge, which, if not frequently removed, becomes fetid. In marked cases, little red nodules in clusters grow on the affected limb, and these are termed grapes. This latter condition is not uncommon among heavy cart horses, which have been kept in dirty stables, and are not well attended to.

As grease is a condition which only too often becomes chronic, leading to permanent inflammatory thickening of the limb, one cannot afford to neglect the treatment of this unsightly affection. In the first place, in these cases, it is well to commence treatment by the administration of a moderate dose of aloes, say four or five drachms. When it is not possible to rest the animal, of course it is not advisable to administer aloes. Whether the physic be given or not, one ounce of bicarbonate of potassium, with two drachms of iodide of potassium, may be given in the drinking water or in the food, twice daily for a week. In many instances this will usually prove sufficient to effect a cure. At the same time the boracic acid ointment, or the ointment of oxide of zinc, or the compound petroleum ointment may be applied locally to the affected parts. It is our practice to give all our horses one bran mash weekly. In cases of grease, it is well to give two at least. Should the above treatment not prove curative, we may administer two drachms of Fowler's solution, with six drachms of bicarbonate of potassium every morning and night for a week, in the drinking water. In the middle of the day, balls made of calomel one drachm, with ginger, gentian, and treacle to eight drachms, may be given in addition twice weekly.

When grapes are present, it is our custom to burn them off with the actual cautery. When grease is due to insufficient exercise, this deficiency should be remedied. In spite of all treatment, some cases of grapes are very inveterate, and in such cases bichromate of potassium has been given internally, but without much success. The causes on which grease depends, are very similar to those which give rise to eczema, and it will be observed that the treatment of the two affections does not essentially differ. Of all the internal remedies which it is our custom to administer in the treatment of grease, we have found no one to be of greater value than iodide of potassium in two drachm doses. This preparation we may give as we mentioned above, in the drinking water or in the food, or in the form of a drench. One ounce of bicarbonate of potassium is a valuable adjunct, and, in pronounced cases, two drachms of Fowler's solution may also be given. This mixture may be administered twice daily. If preferred, iodide of potassium may be given in the form of a ball in one drachm and a half doses, with one drachm of nitrate of potassium, made up to eight drachms with gentian and ginger. These balls might be given twice daily. We may lay stress on the fact that it is well to commence treatment by the administration of a moderate dose of aloes, followed by three full days' rest, and dieting on bran mashes and warm water. Recently, the writer was called in to a hunter suffering from acute laminitis, brought on by working two days after the administration of a full dose of aloes. Fortunately, this case was taken in hand at the outset, and has made a speedy recovery. When an animal is cured of grease, it should be our object to prevent the recurrence of this unsightly affection. With this view, a moderate dose of aloes may be given occasionally, when it is convenient to rest the animal for a time. Bran mashes, to which are added one drachm of nitrate of potassium, and one ounce of bicarbonate of sodium may be given at intervals, as may be necessary. Regular exercise must also be enjoined.

There are many other remedies for grease, but most of them are not very efficacious. When all remedies prove unavailing, it is often advantageous to turn the animal out to graze for a time. Some time ago, we were called to a valuable three-year old stallion suffering from a very severe form of grease. The disease began at the heels, and spread rapidly as far as the hocks of both legs. Very large hard elevations grew quickly, and the fetlocks were soon covered with thickly-crowded masses of these unhealthy excrescences. From them there oozed continually a thick fetid discharge. Internal remedies proved unavailing. The animal was cast, and through the tissues diseased lines were drawn longitudinally with the firing-iron. In about a month the horse was nearly well, though shortly afterwards he lost flesh, and his appetite failed. He was then turned out to graze, and soon recovered his strength. The legs also greatly improved. Unfortunately, since the horse has come up, the affection has again broken out. In this case, so inveterate is the disease, that when cured for a time it breaks out again, and remedies. prove valueless. Just lately, we have had two similar cases in *older* animals. These have been treated similarly with firing and the administration of internal remedies, and are now cured.

MALLENDERS AND SALLENDERS.

THE affection termed mallenders, when the skin behind the knees is attacked, and sallenders, when the integument on part of the hocks is involved, is a similar disease to that scaly condition of the skin on the human being, which goes by the name of psoriasis. Psoriasis also sometimes affects the integument of other parts of the horse, and is not very uncommonly seen in front of the withers. It is a difficult affection to cure radically, and for this reason it is regarded as constituting unsoundness. Of its causes we know very little indeed, but overfeeding and heredity both seem to play their part in its production.

In these cases the diet should be laxative, and limited in amount. It is well to commence treatment by the administration of a moderate dose of physic. The patches should (when it is feasible) be poulticed, to remove the scales, and they should then be anointed twice daily with ointment, made of chrysophanic acid half a drachm, and benzoated lard one ounce. Internally, two drachms of Fowler's solution, with six drachms of bicarbonate of potassium, may be given twice daily, in the drinking water, for a fortnight. This treatment may then be discontinued for a week or so, and then, if necessary, may be again resumed for a similar period.

MUD FEVER.

MUD fever is an inflammatory condition of the skin of the limbs, of a nature similar to that of cracked heels. It depends upon similar causes, being commonly brought on by washing the limbs in hot water after a day's work. It may also be brought on after work in wet weather, by the irritating action of the mud alone. Sometimes the inflammatory condition invades the skin of the belly also, and in almost all cases, the integument of the horse being very sensitive and irritable, it is not to be wondered at that a certain amount of febrile disturbance is manifested, and the animal is rendered unfit for work for a time. We have insisted above, that the plan of removing mud by washing with water was to be reprehended, and more especially that the use of hot water was to be condemned ; and we may now repeat the injunction; for mud fever rarely or never appears when the mud is allowed to dry on the limbs. When the limbs of the animal have been stripped of hair by singeing or clipping, inflammatory conditions of the skin, such as mud fever, cracked heels and grease, are much more likely to follow the practice of washing, and this is especially the case when the parts are not afterwards thoroughly dried. Washing the legs with warm water is never a good practice, but with cold water evil effects rarely follow, if the limbs are dried thoroughly and completely. It is the cold or chill produced by evaporation which stops the action of the skin, and leads to an irritable condition of the part.

We may conclude our remarks on this subject by adding that, when washing is practised, it should be done with warm water, which should not be hotter than about 70° or 80° F. The limbs should be dried, and after applying a little almond or other oil, they should be carefully bandaged. Now, with regard to treatment, the disease, being very similar to cracked heels, demands similar remedies. It is well to commence by giving a moderate dose of physic, and resting the animal for three days ; in the meantime, feeding him on bran mashes and warm water, with half an ounce of bicarbonate of potassium in it, twice or three times daily. When it is not possible to rest the animal, a pint of linseed oil may be given instead of the aloes. Locally, we may use cooling lotions or soothing ointments to allay the irritable condition of the integument. A good ointment may be made of four parts of vaseline to. one of glycerine ; or the compound petroleum ointment may be employed, when there is much soreness. A good lotion is made of half an ounce of liquor plumbi subacetatis, half an ounce of methylated spirit, and seven ounces of water.

WARTS.

WARTS are mostly met with in the horse in those parts where the skin is thinnest, and most abundantly supplied with sensation. The lips, nostrils, eyelids, the lower part of the belly, the sheath, and the udder, are more frequently the seat of warty growths than other parts. Unless warts interfere with the general usefulness of the horse, they cannot be regarded as constituting unsoundness, except when they are so abundant as to prove very unsightly.

There are various surgical methods by means of which warts may be removed. These include ligaturing with thread or silk, burning with the actual cautery, cutting off with the knife, and, finally, the use of caustics. It should be pointed out that by whatsoever method warts are removed, care should be taken to remove the whole of the growth, since otherwise a recurrence of the excrescences may be expected. Ligaturing is an easy method of eradicating warts, where the attached ends do not cover a much larger area than the apices. Ordinary thread, silk, or horse hair should be very tightly tied around the growth, thus stopping its nutrition. When the wart has fallen off, the surface may be touched with a stick of lunar caustic, and any fresh excrescences may be thus treated, should they appear on the site of removal.

When the warts have a large area of attachment, they are most easily removed by the knife, after which the bleeding may be stopped by mopping the surface with cotton wool or tow, dipped in tincture of perchloride of iron. Should any new excrescenses appear, they may be treated as mentioned above with a stick of lunar caustic. Sometimes it is most convenient to remove warts by means of strong caustics, or by the actual cautery. One of the strongest caustic mixtures is the "arsenic paste," but this must be used with caution, and is only adapted for treating warts not situated near any very sensitive organ, as the eye or nose. Caustic potash is sometimes used with a similar object, but it is difficult to keep it from running over the healthy skin around. Burning warts is sometimes the best method, and when judiciously performed, it causes very little or no pain.

Finally, when the stalk of the wart is very thin, a piece of string may be tied round the root, and then by pulling forcibly, the whole growth may be enucleated. Some months ago, we removed about one hundred and twenty warty growths from the nose and face of a two-year-old. Some of these were very large, others smaller. These growths we removed with the knife, and the bleeding was controlled by the use of the tincture of perchloride of iron.

Our readers will understand that our remarks on warts merely apply to those cutaneous horny excrescenses, and not to tumours or lumps embedded beneath the skin.

CHAPTER II.

DISEASES OF THE FEET.

General remarks on the Anatomy of the Foot. Laminitis; Acute, Sub-acute, and Chronic. Navicular Disease. Sand Crack. Canker. Thrush. False Quarter. Corns. Seedy Toe. Quittor. Tread. Over-reach. Villitis, or Inflammation of the Coronary Band. Carbuncle of the Coronary Band. Horn Tumours. Pricks and Injuries of the Foot. Side Bone. Ring Bone.

ANATOMY OF THE FOOT.

BEFORE commencing our account of the numerous and important diseases of the horse's foot, it is our intention to give our readers some account of the structures contained within the hoof, and then to describe very shortly the horny covering itself.

It is quite impossible for anyone to have accurate views regarding the nature and treatment of the various diseases of the feet of the horse, unless he first makes himself acquainted with the main facts regarding the conformation of these marvellously constructed organs of progression.

Most of our readers will be aware that the so-called knees of the horse correspond, not with the knees or elbows of man, but with his wrists. These joints of the horse, like the wrists of man, are made up of a number of small solid bones. There extends from each wrist joint of the horse one long bone, called the shank or canon bone, on each side of which is placed one rudimentary bony appendage, termed a splint bone. Now this canon bone corresponds with that bone of the human hand which extends from the wrist to the root of the middle finger, and it is rounded at its extremity, where it enters into the formation of the fetlock joint. The fore and hind fetlock joints thus correspond with the joints at the root of the middle finger and the middle toe respectively of man. Just as there are three digits in the human fingers and toes respectively, so there are in those of the horse. In the accompanying figure (from Chauveau) of the right foreleg of a horse, A is the so-called knee joint; B is the canon bone; C, one of the two splint bones; D, E, and F are the first, second, and third digits respectively. The third digit, F, is commonly termed the coffin bone. X represents the lower surface of the navicular bone.

K

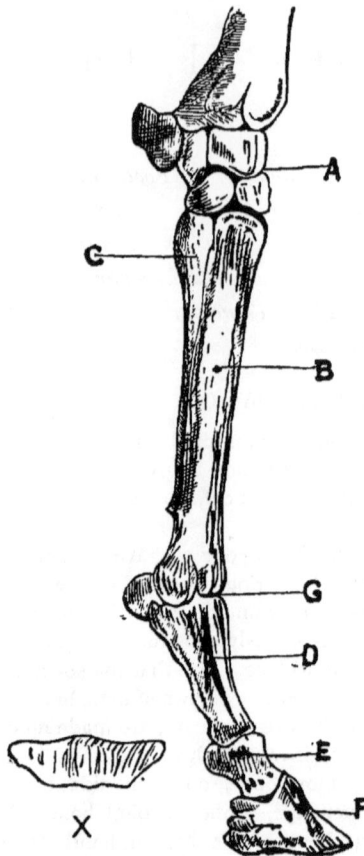

The upper end of the first digit forms with the lower end of the canon bone, the fetlock joint G. The second digit is not so long as the first, and the third one with the lower end of the second, are enclosed in the horny case termed the hoof.

We may pause awhile to inquire what are the two splint bones of each leg? They are rudimentary bones, whose representatives in the progenitors of the horse, were well-formed canon bones. The Hipparion, found in those formations known to geologists as the late Miocene, was a small graceful animal, having three well-developed toes, each bearing a hoof. The middle toe was strong and large, while the lateral ones were so small as not to reach beyond the fetlock. It is noteworthy and most extremely interesting, that cases are recorded where horses have been born with a three-toed foot, in all respects similar to the Hipparion. The earliest ancestral form of the horse was the Eohippus, found in the Utah territory of America. It was of about the same size as a fox, and each of the four feet was provided with three toes.

The horse's hoof contains, the coffin bone ; the lower part of the second digit ; the four ligaments binding the joint between these two digits ; a tendon in front of the joint, which extends the foot ; a tendon at the back of the joint which supports it, and is fastened into the coffin bone behind, after gliding over the back surface of the navicular bone. This tendon flexes the foot when called into action. In addition to these structures, the hoof contains the lateral cartilages of the coffin bone, the matrix or membrane which forms the horny covering, and, lastly, the so-called cushion of the foot.

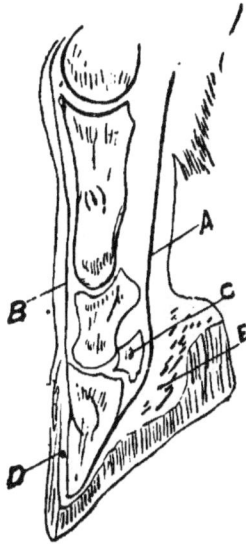

In the above figure of a section through a horse's foot, A is the tendon that bends the foot ; B is the tendon which extends the foot ; C is the navicular bone ; D is the sensitive membrane covering the coffin bone. Its surface is covered with elongated vascular outgrowths, which fit into depressions in the horny covering of the foot ; E is the so-called cushion of the foot.

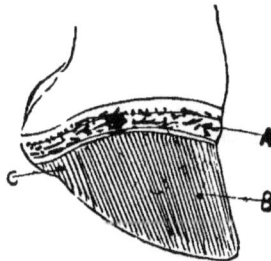

This picture is a lateral view of the horse's foot after removal of the hoof. A is the coronary cushion ; B shows the vascular prolongations ; C is the cushion of the sole of the foot (after Chauveau).

The coffin bone or third digit is peculiar in the horse. Both in structure and economy there is a close analogy between this bone in the horse and the double form of the same bone in the ox; but the resemblance is only partial, each bone being fitted for the special purpose for which it is wanted. The cloven-footed animal moves with astonishing security over granite rocks, where the horse is less adapted to venture. This fact is shown also in the different kinds of goat and deer, and in a lesser degree in the ox. All cloven-footed animals are endowed with wonderful security of foothold, but they lack the elasticity needed to carry weight, as well as the graceful movement of the horse, with his ample security of footing over hill and dale. In such places the noble creature finds sustenance for life, and here his special powers of speed and endurance are required. The coffin bone has much of the form of the hoof in its exterior aspect, and when the lateral cartilage, with the other structures attached to it, is seen in connection the whole structure is similar in its outward form to that of the hoof (Gamgee). On

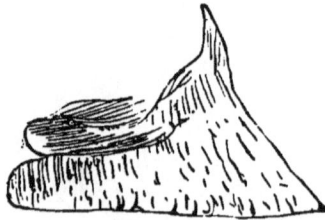

reference to the accompanying picture of the coffin bone, it is seen to resemble in its leading features the external form of the hoof, one chief difference being found to consist in the former being fully a fourth shorter than the inner cavity of the latter. When, however, this wonderfully-constructed bone, of which we append a drawing, is furnished with the cartilages, ligaments, tendons, and all the other important structures, of which it constitutes the centre, it assumes the form and becomes the counterpart of the hoof.

The horse's hoof is not only to be regarded as a covering for the protection of the sensitive structures from injury. The hoof has its specially assigned place in the whole economy of the foot, and each separate component part must be looked upon as an essential constituent of the whole organisation. The hoof forms an integral part of the foot, and those animals that lack it, though amply protected as they are, cannot sustain weight and undergo the same fatigue on the same spots, as those which possess it. Of all creatures gifted with the hoof, the horse is the superior (Gamgee).

The inner face of the hoof presents over its entire extent white parallel leaves, which dovetail with the prolongations or laminæ of sensitive vascular tissues. The hoof has three parts—the wall, the sole, and the frog; and the horn composing these parts differs in composition. That of the wall is denser than that of other parts. The density of the fibres of which the horn is composed, is noticed to become greater as they approach the surface of the wall, the outer layers acting as an efficient protector for the inner structures of the foot. The frog is a triangular horn of a very elastic, fine, tough texture, more pliable than the sole. At its front end or apex, it is seen to consist of a single ridge, but behind it is cleft to allow of motion in the posterior part of the hoof.

The structures of the foot are abundantly supplied with blood vessels and nerve filaments, and owing to their encasement in a horny covering, the foot is generally the seat of acute pain when inflamed or injured.

LAMINITIS, AND PUMICED FOOT.

THERE are two distinct diseases affecting the feet of the horse which cause heat, pain, and lameness. The first of these is termed laminitis, and the second navicular disease. The one may lead to a convex condition of the sole, commonly spoken of as a pumiced foot, while the other leads to contraction of the foot. It will be our endeavour in treating of these diseases to clearly demonstrate the difference between these two affections, not only because they are so constantly confounded, but also because they both assume such a variety of forms.

The terms "founder," "chest founder," "fever in the feet," "fourbure," and "laminitis," are all names for the same disease. The two former terms appear to have arisen from the fact that the horse suffering from fever in the feet, seems to founder or stumble in his gait. Laminitis, or inflammation of the sensitive or vascular structures of the foot may be acute, subacute, or chronic. In fact, it may be seen in all forms, from that manifested by a slow pottering gait, accompanied by no constitutional disturbance, to an acute inflammatory disease, accompanied by high fever, acute pain, accelerated pulse, and high temperature. As a rule, the fore feet are attacked, but not uncommonly the hind ones are also implicated. We agree with Percivall in maintaining that the foot most subject to laminitis is the broad flat one, and that the horses most commonly affected are half-breds and cart-horses. On many occasions however, acute laminitis does attack hunters and thoroughbreds.

Acute laminitis, of all diseases to which the horse is subject, probably causes the most intense agony. The symptoms of this malady are very characteristic. In cases where both fore feet are affected, the action is slow, and the animal seems stiffened, and is often described by the attendants as "fast in the chest." The feet are placed slowly and gingerly to the ground, with the heels first, and the patient "blows hard" as a result of the high fever and great pain. The distressed animal unwittingly makes an effort to move forwards by placing his hind legs as far under his body as possible; he then raises both fore feet, and, as Percivall says, makes a "timid leap forward." In other cases he will gingerly advance first one leg and then the other; and again at other times he will rest in the recumbent posture.   In very bad cases, the horse will slowly rear on his hind legs when made to move.   The pulse is quick and sharp, and the breathing is hurried and distressed.   If the hind feet are alone affected, the fore feet are directed backwards underneath the body.   When all four feet are affected, the patient usually prefers the recumbent posture.   The hoofs when felt are noticed to be much hotter than in health ; and some observers have recorded that they have seen blood ooze from the coronets.   Distinct throbbing of the arteries of the pastern may also sometimes be felt.   The action of a horse affected in the hind feet only, somewhat resembles that of stringhalt.   If the hoof be tapped sharply, the horse manifests considerable uneasiness.   Sometimes in his agony, a horse will perspire profusely and gasp for breath.

The symptoms presented by a horse suffering from acute laminitis, sometimes lead to the supposition on the part of the unskilled, that it is inflammation of the lungs, which the animal is affected with.   We have often been called to bad cases of so-called inflammation of the lungs, which we have found to be in reality, pure and uncomplicated attacks of " fever in the feet."   In this connection, however, it may be mentioned that horses affected with this disease, sometimes suddenly develop inflammation of the lungs or bowels, or pleurisy, or first one and then another.   In pathological language, this kind of rapid change in the seat of a disease is called "metastasis."

During the year 1883, we were called to attend a five-year old valuable dark brown hunter, the property of a dealer resident in a town in Lincolnshire.   The number of respirations per minute was 74, the temperature 105° F., and the pulse 96.   The disease was confined to the fore feet.   As the horse was very plethoric, four quarts of blood were abstracted from the jugular vein.   The animal was bled locally at the coronets.   The feet were afterwards placed in a tub of hot water for over twenty-four hours.

On the following day, the animal was much better, and hot bran poultices were substituted for the hot water.   The pulse was now 82, and the respirations 54.   On the third day, although the feet were better, the pulse was 84, and the breathing numbered 68 per minute.   At this juncture, as frequently happens in acute laminitis, the horse developed all the symptoms of acute pleurisy.   This complication was treated by the application of the hot pack to the chest, the internal remedies not being altered.   On the fourth day, the pulse remained the same, but on the fifth and sixth days it gradually fell to 52.   After this time the patient improved rapidly.   On the seventh day, towards evening, the owner came in a hurry to announce that the animal was colicked.   We found him to be suffering from acute inflammation of the bowels, and the pulse again rose to 80.   This new complication was treated with morphia, and mustard poultices were applied to the belly.   On the eighth day, there was not much alteration in the condition of the patient, but on the ninth, the pain ceased, and the pulse fell.   Then, until the twelfth, the animal continued to improve.   On the thirteenth day, the owner turned the animal out to grass, against orders. On the following day, he came announcing that the animal had got stringhalt. This we found was simply laminitis, now developing in the hind feet. The animal was brought up from grass, and the hind feet were poulticed until the seventeenth day, by which time he was much better. On the eighteenth day, the owner again turned out the animal, thinking him to be suffering from stringhalt, and refusing to believe it was laminitis. He was told that the soles of the feet would come down, but this advice was unheeded. In about a fortnight the animal was worse. We went to see him out at grass, and found that the coffin-bone had appeared through the horny sole of the foot. We thereupon recommended carbolised dressings to be fixed on the feet with leathern boots. Two months afterwards the animal had so far recovered, that he was sold to a gentleman for slow work. The horse now works very well on the land. This case, of course, was a very severe one, and we mention it to show some practical points which it may be useful to know.

We have met with several cases where the soles have come completely down, and the animals have afterwards so far recovered as to be able to work. One, which occurred at an hotel in Louth, was a very similar case to the above-mentioned one, and extended over a period of three weeks and a few days. This animal was for many days at the point of death, and not only suffered severely in all four feet, but, in addition, had extensive lung disease.

The acute symptoms described are not invariably present even in bad cases. Early last year I was called to a bad case of laminitis, affecting the hind feet in a grey twelve-year old hunter. Although in this instance the constitutional symptoms were not severe, yet throughout the disease, they were very persistent. The bones were so extensively diseased in this animal, that there was no prospect of his being able to hunt again, and he was therefore humanely slaughtered.

Cases of acute laminitis, when the constitutional symptoms are fairly severe, are of frequent occurrence among heavy draught horses.

Subacute laminitis is also of frequent occurrence, but is not so sudden in its appearance, or so rapid in its progress as the acute form. The gait of the animal is stiff and "groggy," but the general disturbance is not severe. Like the acute form, this also may lead to chronic laminitis.

Lastly, we turn to the consideration of chronic laminitis, or pumiced foot. This affection may be the sequel of either acute or subacute laminitis, and never has any other origin.

The disease consists in the union between the sensitive or vascular laminæ, which form the horny structure of the foot becoming detached from the hoof. The coffin bone, pressed down by the exudation poured out by the inflammatory action, and losing its ties of suspension, sinks down upon the horny sole. It must be remembered that a horse may have a very flat sole, without necessarily having a pumiced foot.

We were once called to a cart-horse which had been going lame in both

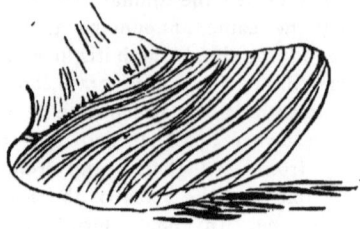

The figure represents a Lateral View of a Foot afflicted with Chronic Laminitis.

fore feet for some months. The gait of the animal clearly indicated the nature of his disease. The shoes were removed and the feet cleaned out, when both soles were found to be convex downwards. This was a pronounced case of pumiced foot, and it had been a slow subacute form of laminitis, resulting in descent of the sole. Such feet should be shod with round, stout, broad-webbed bar shoes.

The above picture is a section of a pumiced foot, showing (A) the coffin bone descending, and (B) the convex sole. This is the intermediary stage. In still more advanced stages, the hoof grows more forward, as shown in the first picture, and the end of the coffin-bone becomes bent forward, and the sole becomes elongated and flattened. (After Signol).

This figure shows the third (advanced) stage of Pumiced Foot.

The predisposing causes of laminitis are:—working animals when out of condition, fast trotting on hard roads, and inherited tendency. Allowing horses to drink cold water when heated, and keeping them in a standing posture for a long time on board ship or in slings, are also causes of this painful disease. Mares in foal are often afflicted with a mild subacute attack of laminitis. An overdose of aloes has been known to cause laminitis ; and this disease, according to Percivall, may in some seasons become epidemic. We have, however, never known of its becoming epidemic. Finally, we may add that barley, wheat, and Indian corn often cause acute inflammation of the feet.

Removing animals suddenly from grass, and then overfeeding them in the stable, is a common cause of the acute form, more especially when the animal is suddenly called upon to perform work. A good instance of this we met with a short time ago. A horse belonging to a carter escaped in the night, and made his way to a bin of powdered Indian corn, of which he devoured over two stones. The owner next morning drove him from Louth to Grimsby and back, a distance of about twenty-eight miles. The following day the horse was struck down with acute laminitis. We were called in when the horse had already been ill about fourteen days, and found the coffin bones in a state of osseous mortification, or necrosis. The animal was accordingly ordered to be shot.

There are few diseases of which such erroneous views are generally held, and contrary to the general impression, there are few which are so amenable to early, judicious, and careful treatment, as is laminitis. If treated properly in the early stages, cases of inflammation of the feet will very often completely recover. It may be stated most emphatically, that the earlier they are attended to, the better is the chance of ultimate cure.

A short time ago I was requested by a large land-owner to examine a valuable cart-horse which was being treated for ringbones. The animal was trotted once down the yard. " Your horse," I observed, " is suffering from chronic laminitis, and must be treated accordingly." Here was a valuable animal, worth ninety pounds or more, well nigh wrecked, and now scarcely worth fifteen pounds, simply because the owner had been misled as to the nature of the disease. Such a case was one which would have been pre-eminently curable in the early stages. In the later ones, when the malady had become chronic, the disease proved much more refractory. This animal made a good recovery eventually, but at a far greater amount of trouble and expense, than would have been necessary, if the case had been attended to properly in the first instance.

In treating acute laminitis, it is our practice to administer three or four drachms of aloes in the first instance, or one pint of linseed oil. We do not, of course, administer any purgative if the bowels are already too freely opened. The diet should be laxative, consisting of bran mashes and linseed cake gruel. The shoes should be removed, but it is not desirable to pare away any of the horn. Internally a drench, composed of five drops of Fleming's tincture of aconite, one ounce of bicarbonate of potassium, and four ounces of liquor ammonii acetatis, may be given every four hours for four times, and then every six hours on the following day. Afterwards, the aconite should be omitted from the draught, which may be given three times daily. The bicarbonate of potassium is greatly to be preferred to the bicarbonate of sodium. In the case of very plethoric animals, three or four quarts of blood should be abstracted from the jugular vein, and in those cases where the arteries of the feet pulsate very distinctly, we may remove a pint of blood from each coronet, by puncturing with the lancet. The feet should be kept in tubs containing water at a temperature of about 110° F. After the first day, poultices of bran may be substituted. When the animal becomes much better, bar shoes shoud be put on ; and the animal should be gently exercised for half an hour daily. The exercise should be gradually increased, until recovery is complete. After the acute symptoms have abated, it is customary to have the coronets well blistered.

In cases of chronic laminitis, the soles or frogs should not be pared on any consideration ; but the animal should be shod with leathern plates, upon which the bar shoes should be nailed. The coronets should be blistered with red iodide of mercury ointment. This blistering may be repeated in three or four days, with equal parts of lard and red iodide of mercury ointment. Pressure upon the frog is important in shoeing in this condition of the foot. In those cases where the animal manifests uneasiness after work, or when the hoofs are abnormally heated, it is well to allow the horse to stand with the fore feet in cold water for an hour or two. Where this is unnecessary, the horse is better on a cool than on a heated bedding.

NAVICULAR DISEASE AND CONTRACTED FOOT.

THERE is no more important subject in the whole range of veterinary surgery, than the one which it is our intention now to discuss. It is not our purpose to enter fully into the many theories which have been propounded concerning the pathology of this obdurate and common cause of unsoundness, which is sometimes spoken of as " groggy," and was formerly known as coffin-bone-lameness. Indeed, with the small space at our disposal this would be absolutely impossible, for one might write an elaborate treatise on this subject. It is rather our object here to draw the notice of horsemen to .those practical facts which, from every point of view, deserve to receive very careful attention.

The navicular is a transversely elongated bone situated at the back of the coffin bone. It is flattened above and below, and narrowed at the extremities. It is made up of an outer layer of dense compact bone, enclosing very dense spongy bone. Behind this bone there is a very important tendon, which passes round it to become attached to the back of the coffin bone. This tendon passes upwards behind the limb, and joins the muscle which, when called into action, bends the foot.

Navicular disease arises in the first instance from inflammation of the navicular bone, or of the cartilage on the under surface of this bone. This inflammation often spreads to the thin lubricating membrane between the bone and the tendon, and then attacks the tendon itself. At length, as the disease progresses, these structures become welded together, and weakened by the products of inflammatory action.

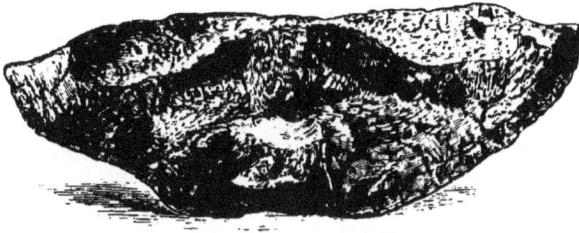

The accompanying picture is engraved from a photograph of the lower surface of a diseased navicular bone, one-eighth enlarged. It shows the caries of the bone.

The picture below (Sewell) represents a section of the horse's foot, showing (A) the navicular bone, (B) the flexor tendon, and (C) the coffin bone. It will be noticed that the navicular bone forms a kind of pulley round which the tendon works. We shall have occasion to refer to this picture again, as showing the method of treating navicular disease by the operation of frog-setoning.

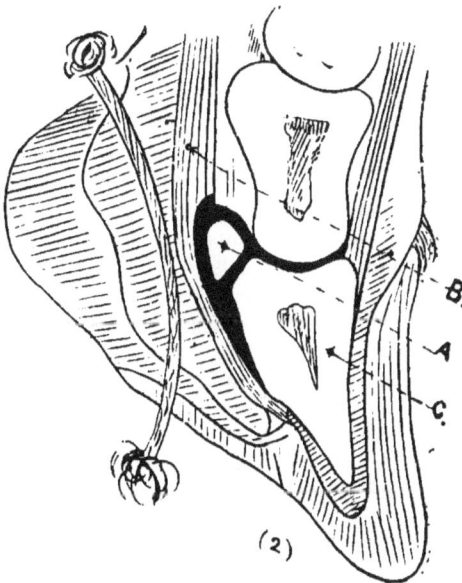

(2)

We shall first speak of the causes to which navicular disease is to be attributed; then of its symptoms and methods of detection, in this connection dealing with contracted foot; and, lastly, we shall review some of the various methods of treatment, unfortunately in many instances so futile.

Hereditary influence, that potent predisposing factor in the causation of so many diseases, is often clearly traceable in navicular disease. The practical conclusion to be deduced from this fact is, that animals afflicted with this malady should not be used for breeding purposes. It has been said, and we believe with good reason, that feet with high heels are more liable to navicular disease than open flat ones. When the soles are flat, there is necessarily more constant pressure on the frog, which, together with the structure it supports, is thus maintained in a more healthy state. It must be borne in mind that heredity is probably of influence not so much in predisposing animals to inflammation of the navicular bone, as in transmitting to the offspring that peculiar shape of foot which is especially liable to lead to such changes in this important structure. In any case the knowledge that heredity plays a great part in the production of this disease is of value, as indicating that affected animals should not be used for stud purposes.

When we consider that the hind feet are very rarely affected with navicular disease, but that the fore feet are very commonly so diseased, one naturally expects to find some cause which, though very rarely resulting in changes of this kind, is very potent in leading to disastrous lesions in the fore feet. What is the cause? It has often been noticed that this affection is far more commonly met with in horses used for quick work on hard roads, than among other animals. We naturally infer from these facts that navicular disease is largely dependent upon concussion or undue jarring, which necessarily affects the fore more than the hind feet, and is more violent and sustained in roadsters and hacks.

The disease, it may be stated, often begins in the membrane lining the navicular joint, or in the cartilage lining the surface of the bone, which is the spot where we should expect that the effects of constant and violent concussion would be especially liable to result in inflammatory changes.

Among hunters and racers, navicular disease is not nearly so common; for their work on softer ground does not cause this violent jarring. It is to the quick, long journeys made on the hard roads that this inflammation leading to such disastrous consequences is mainly due. Regarding these points Mr. Stewart long since wrote, "long journeys, performed quickly, will make almost any horse 'groggy.' Bad shoeing and want of proper care also help, but alone they never produce this affection. The animal must journey far and fast; but, if his feet be neglected, or the shoeing be bad, a slower pace and a shorter distance will produce the mischief."

As we might expect, navicular disease is almost unknown among cart-horses; but they are more subject to laminitis than finer bred animals. In addition to the major causes of navicular disease, there are some minor ones we may shortly mention. Rheumatism is believed by some to be a cause of navicular disease, but we do not think this is ever the exciting factor.

Rarely, navicular disease may be due to injury of the foot caused by nails or bruises. Finally, slanting pavements are believed by some to favour the production of disease of this bone.

When lameness comes on gradually in a horse six or seven years old or more, and the animal points his foot in the stable, we have strong grounds for suspecting commencing navicular disease. It is of importance to be able to recognise this affection in its early stages, before the disease becomes chronic ; for not unfrequently therapeutic measures may be taken to prevent the progress of the inflammation. This form of lameness, although usually affecting animals of six years old and upwards, is not very uncommon in younger horses from three to four years old. When the disease first begins, there may be but little to attract attention beyond the pointing of one or both fore feet in the stable, an abnormal warmth of the hoofs, and a scarcely perceptible lameness, perhaps only manifested at times, and disappearing after exercise. As the disease progresses, the lameness increases, and is more marked after rest ; especially when this is preceded by a journey of seven or eight miles sharp trotting on hard ground. If the foot be examined, it is sometimes found to be hotter than normally, and as a result of the disease of the navicular bone it becomes contracted ; but it may be pointed out that contraction is not always present in navicular disease, nor is every contracted foot necessarily accompanied by this affection of the bone.

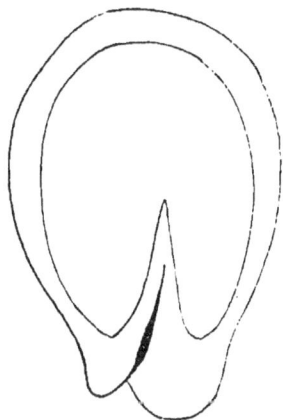

CONTRACTED FOOT.

Unless the case be somewhat advanced, the animal generally walks sound ; but betrays his disease by his short groggy steps when trotted, especially when going at a sharp pace over hard stones. The horse digs his toes in the ground in order to obviate the pain, which would be caused by bringing the heels firmly down. The iron at the toes of one or both shoes becomes worn away in consequence. The habit of pointing the fore feet, which is done by the animal to ease pressure on the heel, is a characteristic sign of navicular disease. When an animal thus affected is to be sold at a fair, he is commonly tied up closely to the manger, so that he cannot point his feet, and thus the unwary are deceived ; for the lameness may not be apparent in the cursory trot up and down the soft ground outside the stable. The best way to detect this insidious malady, is to ride the animal six or eight miles briskly, and then let him stand *loosely tied*. If affected with navicular disease, he will then probably soon point one or both fore feet. After resting a quarter of an hour, he should then be led out of the stable, and trotted up and down on hard ground. The peculiar characteristic gait will then in most instances become apparent. In these trials the animal should carry a good fair weight.

Horses affected with navicular disease generally stumble a great deal, and thus not uncommonly break their knees ; but, when worked judiciously, and not trotted fast on hard ground, they may work well for many years. On soft ground, affected animals are benefited by regular work, and may be used for hunting or other purposes. Indeed, it is not at all uncommon to see a horse with navicular disease in the chase. When the disease becomes confirmed, the lamenesss does not necessarily increase : for the caries of the bone and its cartilages may remain in a somewhat similar condition for years. In most cases of navicular disease, the lameness is most pronounced on first leaving the stable ; but it gradually disappears, perhaps entirely, during exercise.

The treatment of navicular disease will necessarily vary considerably with the nature of the case. In those very acute cases which come on very suddenly in the course of a week or so, it is advisable to give the animal five or six drachms of aloes, and feed him for three days on warm water and bran mashes. Poultices for a week or a fortnight will also prove very useful. Internally one ounce of bicarbonate of potassium may be given twice daily in the water for a week or more. Blistering the coronets with ointment of biniodide of mercury, and the turning out the animal to grass for six months will sometimes effect a cure. Bleeding from the coronets is

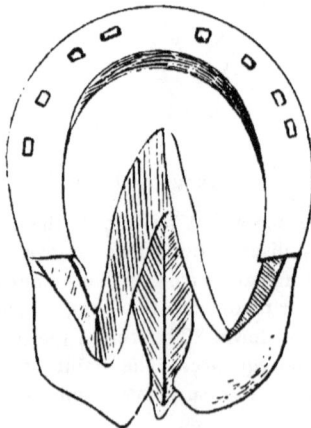

in all instances worse than useless in the treatment of any form of navicular disease. In ordinary chronic cases of navicular, if the animal can be rested for six months, he should be frog-setoned and then turned out to grass. Blistering may be adopted instead of frog-setoning. If he cannot be rested or if the disease is not so marked as to necessitate cessation from work, it is best to have the animal lightly shod, and to apply swabs moistened with cold water. The work should be gentle. Hacking is the most suitable of all kinds of work.

The heels should be rasped down a little. Half-moon shoes as represented, may be applied, or we may adopt the Charlier method of shoeing. We append a representation of this method of shoeing from M. Signol's "Aide-mémoire du Vétérinaire."

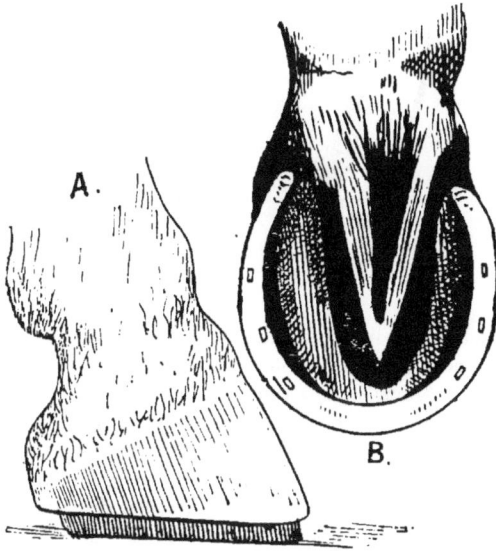

Veterinarians have devised many forms of shoe for the alleviation of chronic navicular disease. We append two of these, but we cannot speak of their value, as we have no experience of their efficacy in preventing the foot from becoming contracted.

Toe clips should be very small, or dispensed with altogether. The heels of the shoe should be thickened. The "Thacker" shoes are strongly recommended by some. We may here remark that large toe clips are not uncommonly a cause of disease of the foot, and should never be used.

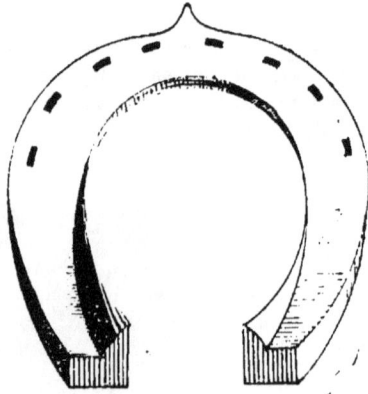

Animals subject to navicular disease, require a cool bedding on a level pavement, but nothing can serve the purpose better than sawdust, or a good supply of straw. Animals, as a rule, will lie down in a quiet well-bedded box more readily; and, as the recumbent posture is to be encouraged at night, it will be well to make the box as comfortable as possible.

When these measures have been taken, there remain two operations which have been devised for the cure of navicular disease, and of these we shall say a few words. The first, frog-setoning, as represented in the second of our illustrations, is said to have proved serviceable in some inveterate cases, when all other measures had failed. The second operation for the alleviation of navicular lameness is termed neurotomy. It was at one time so highly thought of as to be very frequently performed. It consists in removing a portion of nerve from both sides of the plantar nerves of each fore limb. It is sometimes a very successful operation. It is, however, not advisable to perform it except *as a last resource, when all other measures have failed, and the animal is quite unfit for work;* as, although it often affords temporary relief, the nerves usually grow together again after a time, and the animal may become still more lame than before. In performing the operation, it is usual to remove about one inch from the nerve of each side of both fore limbs; but sometimes it is performed on one limb only. Neurotomy was supposed to have been first introduced by Mr. Sewell, but we have reason to believe that the operation was practised some years earlier by Mr. Moorcroft.

After division of the nerves, the part below the seat of section loses sensory power, and, no pain being felt, the animal often ceases to manifest lameness.

It is scarcely ever advisable to perform neurotomy on a young horse. Some time ago, we performed the operation on a four-year-old colt, on which all other methods had been tried. The animal has since been perfectly sound, having made a good recovery. Our readers must bear in mind that the treatment of navicular disease is at best mainly unsatisfactory. It is always well to dispose of animals so affected, when an opportunity offers itself.

SANDCRACK.

By a sandcrack we understand a longitudinal fissure of greater or less extent in the horny fibres of any part of the wall of the hoof, commencing close to the coronet, and mostly found at the inner quarters of the fore feet and at the toes of the hind feet. More rarely these fissures are met with in the outer quarters. At first the fissure is small, but it gradually extends downwards and upwards. Wherever situated, sandcrack constitutes unsoundness.

In the horny horn, the fibres are held together by an agglutinating cellular substance and therefore they do not become separated. When, however, the secreting membrane is injured by concussion or other cause,

L

the fibres become separated, and a crack or split is formed, as the result of the impaired secretory action of the injured part. These cracks nearly always come on gradually. Indeed, the writers have rarely known a crack to be formed suddenly. Those fissures recorded as having arisen suddenly, are in reality to be put down to constitutional disease of old standing. The brittle condition of the crust, altering the character of the horn secreted, it will thus be seen, is to be attributed to an unhealthy action of the membrane secreting the horny fibres, and of the substance binding them together.

In some animals, the horn is more brittle and weak than it should be, and is more liable to crack on any unusual strain. Such a condition of horn, more commonly met with in animals bred in damp, low-lying districts, though not unfrequently inherited, is no doubt in many instances traceable to badly devised methods of shoeing. When the sole and frog are made unduly thin by paring, and seated shoes are used, the weight of the animal is thrown on the crust of the wall only, instead of being more uniformly distributed ; and, as a consequence of this, the membrane secreting the horn is liable to suffer, when the crust it forms is subjected to additional strain. Naturally, this will be especially liable to occur when the badly-shod feet are subjected to continued concussion, by fast riding or driving on hard ground. The fissures are more likely to occur at the inner quarters of the fore feet, and at the toes of the hind ones ; for these are the parts more especially subjected to strain.

As we mentioned, the cracks, at first insignificant, lengthen and deepen, and thus they gradually spread through the horn to the sensitive structures, which become inflamed and bulge through the apertures of the wound. Lameness now becomes a marked feature, and the affected part becomes very painful. Lameness appears before the fissure is evident from the outside, when the crack commences beneath the outer portion of the crust, and then spreads outwards ; and it may be added that it is in such cases as these, that cracks are said to be suddenly made. In reality they have been forming for some time past.

Lameness is more marked when the crack is at the toe, than when in the quarter of the foot, and when involving the toe of the hind foot, it is still more aggravated. As the animal raises its foot, the walls of the fissure widen, and as it places its foot down again, they become approximated, thus pinching tightly the protruding, inflamed tissue, and causing great agony. Sand and mud find their way into the wound, and increase the inflammation and irritation, and at the same time tend to increase the extent of the fissure.

The treatment of sandcrack varies according to the nature of the case. If there be no suppuration, it is our custom to make a horizontal incision with a firing-iron, about one eighth of an inch deep, above the upper extremity of the crack. The crack itself we then fill up with gutta percha. In simple cases, a cure is generally easily effected by these means.

In those cases where there is lameness and inflammation, it will be necessary to remove the shoes and administer a dose of aloes, say four or five drachms, and allow total rest for several days, on a diet of warm water and bran mashes. During this time, the edges of the crack should be carefully pared, so as to allow of the escape of all irritating mud, and foul matter. When this is washed away, the inflamed sensitive tissue beneath may become visible. It will be advisable to have the foot poulticed for a day or two after this operation.

When by these measures the inflammation and consequent pain have in a few days subsided, it is our object to promote the growth of new healthy horn to fill up the crack from beneath ; for the edges of the wound cannot become structurally united. Bar-shoes of a fair thickness may now be applied, and care must be taken to remove pressure from that part of the foot, immediately *opposi.e to* the fissure. When the crack extends as high as the coronet, it is customary to pare away a groove between the upper end of the fissure and the coronet, so as to divide the fissured horn from the substance which forms the new horn. A leathern strap or tarred twine may now be applied tightly round the hoof if necessary, or a clasp may be used. There are two operations which may be performed for the cure of sandcrack. The first, only performed in very bad cases of long standing, is applicable when the cracks extend as high as the coronet. It is termed the stripping method, and consists in making two grooves, each beginning at the coronet about half an inch to an inch on either side of the upper end of the fissure, and joining together at the other extremity of the fissure at an angle of about 70°. The whole of the horn enclosed within this V-shaped area is then stripped off, and the part exposed is protected by the application of a strap.

In those cases where the fissure extends below the junction of the grooves, the horn may be pared away below it also. The animal must now be rested until the gap is filled with newly secreted horn. In the meantime, the foot should be carefully bandaged and dressed with some mild astringent lotion. The coronet may bo blistered so as to stimulate it to increased secreting activity, with an ointment of one part of biniodide of mercury and two of lard. The time taken for a sandcrack to grow up completely from the base, is about nine or ten months. The clasping method is in most instances greatly to be preferred to the above, as being of a far less serious and lengthy character. The operation may be performed in two ways. The French perform the clasping method by burning two holes at equal distances on each side of the fissure by the instrument A raised to a red heat, and then fitting the two ends of the clasp B into them. The clasp when inserted is made tight by pincers, as represented in figure C.

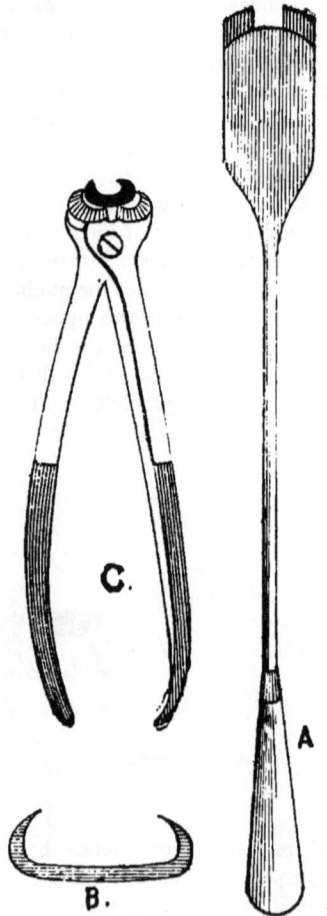

C.

B.

A

(After Signol).

It is best to use a number of clasps placed at a distance of half an inch apart, and when this operation is completed, the foot may be supported by a firm leathern strap. When the horn is thick, more especially when the fissure is seated at the toe, the nailing method of clasping is very useful. A notch is cut about half an inch on each side of the crack, about a quarter of an inch in depth. Several other notches may also be made at intervals of about an inch, if the crack is a long one. Then, by the aid of a skilled smith, horse nails are driven from one side to the other, and their parts are drawn together tightly by pincers, and the ends rasped down. When situated at the toe, a hole may be bored under the crack, and a nail passed through and similarly clenched. When the crack is cured, it is necessary to use flat shoes, and to remember never to allow the smith to thin the sole.

CANKER.

CANKER is a disease characterised by the abundant discharge of thin fetid matter from the frog and sole of the foot, and by the presence of large fungoid granulations, or pallid irregularly shaped elevations, occupying the place where healthy horn should grow. When examined with the microscope, these elevations appear to consist chiefly of imperfectly formed horn cells ; and this would lead us to infer that the horny matter itself was improperly secreted, owing to abnormal changes in the membrane.

Canker most probably depends upon inflammation, and consequent alterations in the membrane, which secretes the horny sole and frog, and covers the coffin bone. The disease usually commences in the frog, extending to the sole, and sometimes involving the sensitive laminæ, which secrete the inner part of the wall of the hoof. Sometimes, the diseased action is confined to one foot, but in other cases it affects two ; and the writer has not unfrequently met with cases in which all four feet were involved. The hind feet are more frequently affected than the fore ones. Canker is very rarely seen except in cart-horses, in which it is not at all an uncommon disease. Regarding the causes of this affection of the foot, it has been suggested by Percivall, that some horses, more especially bulky animals of sluggish lymphatic temperament, are peculiarly predisposed to become affected ; and it is in such animals that the affection termed grease is also especially liable to appear.

Sometimes, canker has its origin in a neglected injury to the foot, in which case it will be confined to the wounded member. Not uncommonly it is traceable to standing on damp and filthy bedding, and to generally bad sanitation.

In canker the horny sole of the foot becomes gradually separated from the membrane which secretes the horn, and, as the unhealthy action spreads, the whole of the sole is thus undermined. Canker is a very difficult disease to treat successfully, and it is therefore advisable to call in the best professional aid. In severe forms, the operation of cutting away the sole is generally necessary. We shall not describe the operation, but we may mention that it consists in taking away the whole of the horny sole of the foot, and the unhealthy growths by which it is undermined. Afterwards, the exposed surface is dressed with some caustic solution, and after filling up the excavation with carbolised tow, or tow saturated with tar, the foot is encased in a leathern boot. A short time ago, we had under our care a very severe case of canker, affecting both fore feet of an aged cart-horse. The animal was totally unfit for work, but it was decided not to perform the operation of stripping the sole. At first the soles were well pared, and the diseased growth was treated by the application of the actual cautery. A week later, the affected part was dressed with the acid nitrate of mercury, and this was renewed every third day for four times. The animal has made a gradual and almost complete recovery, one foot being quite healthy, while the other is progressing very favourably.

One of the best applications for cases of canker is a mixture of four parts of glycerine, to one of pure carbolic acid. Strong solutions of sulphate of copper are also useful.

THRUSH.

By thrush we understand disease of the sensitive frog, accompanied with the discharge of an acrid, foul-smelling fluid, from this part of the foot. In severe cases, the disease spreads between the sensitive frog and the horn, thus causing separation of the latter. The cleft of the frog is the part usually first affected, but the disease, if not cured, may soon involve the whole of this structure.

Thrush may owe its origin to dampness of the ground on which the animal stands, when turned out into low-lying pastures, or placed in damp, ill-drained stables. Not unfrequently it is due to a filthy condition of the litter of the bedding, or to stopping the feet with decayed matter- a common, but pernicious and absurd custom.

It is really wonderful how difficult it is to uproot customs, which by constant use have become so ingrained on the mind, as to be regarded as being beyond question of material value. The practice of stopping the feet with decaying matter, is still a common though most absurd custom. As we said above, the dampness causes maceration of the frog, and by thus denuding the sensitive part of the structure, leads to an abnormal state, which necessarily becomes still more aggravated by the uncleanly matter. In such a way thrush is not uncommonly developed ; but it is fortunate that this condition is generally easily remedied by judicious attention and care.

Dampness causes maceration of the frog, and, by thus denuding the sensitive part of this structure, leads to an unhealthy condition, which becomes still more aggravated, when filth is an additional factor. Want of pressure on the frog is also sometimes a cause of thrush. Lastly, this affection may sometimes owe its origin to constitutional causes, and to frost bite. Lameness is sometimes traceable to this diseased condition of the frog, which necessarily constitutes unsoundness, so long as it remains uncured. Unlike canker, thrush is in most instances easily cured. In the first instance, the animal should have a good dry litter, and the frog should be kept in a clean condition. On no account must the animal be allowed to stand on bedding saturated with excreta, or with accumulation of decomposing matter. Stoppings for the feet must not be employed. The diseased portion of the frog should be removed with the knife, and the affected part dressed once or twice daily with about half a teaspoonful of powder, composed of equal parts of calomel or iodoform, or this powder may be alternately used with one of equal parts of starch and iodoform. A mixture of one part of carbolic acid and four parts of glycerine, is also a very valuable application. Ointment of salicylic acid will also prove of great efficacy.

It is also well to maintain a firm pressure on the frog. In severe cases of thrush, causing lameness, or when there is a tendency to grease, indicating a possible constitutional factor in the production of the disease ; it will be best to commence treatment by the administration of three or four drachms of aloes, followed by three days' rest, during which time the animal should be fed on warm water and mashes.

FALSE QUARTER.

By false quarter we understand the existence of one or more clefts, or deficiencies of horn, in any part of the wall of the foot. Referring to the anatomy of the horse's foot, our readers will remember that the outer horny covering of the wall of the foot is secreted by the coronary substance ; and we may here mention that these clefts are due to destruction of this coronary substance by injury, such as a tread. False quarter is totally different in nature 'from sandcrack. It consists actually in longitudinal flaws in the outer covering of the horn of the wall of the foot ; and at the bottom of the fissures, we find the horny laminæ which are secreted by the sensitive ones. ' Although not usually causing lameness, nevertheless false quarter constitutes unsoundness, as it is liable to affect progression at almost any time, from injury to the thin horny covering of the affected part which is exposed. In cases of false quarter due to recent injury, the affected part of the coronet should be carefully treated. After bringing the injured surfaces together, and applying some antiseptic ointment, such as borax ointment, constant pressure should be applied by means of a bandage.

In old cases, all that can be done is to apply a blister of red iodide of mercury ointment round the coronet, and to fill up the gaps with gutta percha, moulded in, while warm. In addition, the feet may be shod with bar shoes, so as to distribute the pressure more evenly.

Not uncommonly, horses with false quarter are passed off on the unwary, by thus filling up the gaps with gutta percha, and painting the hoof with lamp black or hoof ointment. In order, therefore, to be on one's guard against such tricks, it is advisable, before examining an animal, to have his feet cleaned.

CORNS.

CORNS are bruises or contusions of the sensitive membrane, which covers the lower surface of the coffin-bone, and secretes the horny sole. A corn appears as a small reddish spot or patch, in the space between the bars and the wall at the heel. Corns are almost always met with in the fore feet, though the hind ones are also sometimes affected. In nearly all cases it is the insides of the feet, which are the seats of these bruises, and this is probably attributable to the fact that more weight is thrown on the inner than on the outer side of the foot. We mentioned, in treating of navicular disease, that the fore feet were much more liable to suffer from continued concussion than the hind ones. This would also account for the much more common occurrence of corns in the fore feet than in the hind ones. As in navicular disease, so in the case of corns, it has been observed that animals subjected to constant work on hard ground, are more liable to become affected ; and this is especially the case with high-stepping animals, with weak heels and marked "heel action." In the accompanying picture of the near fore foot, A shows the position of corns between the bar B and the wall at the heel.

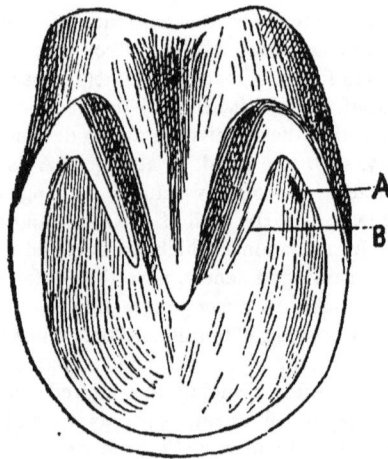

The chief cause of corns is the irrational method of shoeing, which causes pressure at the seat indicated at A in the above picture. In the opinion of Professor Williams, "the ordinary seated shoe is the most irrational and insensate one which ever emanated from man's brain. It is a mechanism which bears upon no part of the sole, except upon the spot which is incapable of pressure. It is dished out, made concave all round the foot except at the heel; and corns result." The seat of corns is just that part where the horny sole is thinnest, and consequently most liable to injury. If the bars have been cut down, and the heels allowed to grow unduly long, corns are more likely to be produced. It must, however, be mentioned that corns are not unfrequently met with in feet in all other respects healthy; but in most instances they only appear as the result of defective methods of shoeing. When the sole is very weak, or has been unduly thinned, corns are naturally more liable to be produced. Sometimes owing to a space left between the shoes and the horny heel, dirt insinuates itself, and pressing on the seat of corn causes the appearance of these bruises. We have already exposed that most pernicious and barbarous custom of stopping horses' feet with decaying matter; and we only allude to it again, to state that by macerating and weakening the horny sole of the foot, it thus renders it far more liable to be injuriously affected by bad shoeing or fast trotting on hard ground, or by any other direct cause. "Stopping" feet is therefore an indirect cause of corns. It is, we wish to point out, not merely owing to the fitting on of the seated shoe, which we have said is so frequently the cause of corns, but also to other mistakes which the smith commonly makes, that these bruises make their appearance. Not uncommonly he pares away the bars, and by this practice, the foot tends to become contracted; and the pressure of the heels of the shoe falls upon the spot indicated as the seat of corn. Lastly, we may add that the use of calkins, and the practice of not renewing the shoes often enough, are to be regarded also as occasional factors in the production of corns. When a shoe is not removed as often as is necessary, and is on the contrary allowed to wear down, it may be removed from its original position, and press upon the seat of corn.

A corn constitutes unsoundness, because, although it may not cause lameness in all instances, or at all times affect progression, yet, until cured, the animal may become so much worse, as to be wholly unfit for work. Rest for several days will often render the horse free from lameness for a time. As a rule, there will not be much difficulty in diagnosing a case of lameness when dependent upon corns. When the horn at the seat of a corn is pared away by the smith, a reddened patch becomes visible, and renders the diagnosis certain. In some cases—and these are not uncommon—all that can be discovered, besides the manifest lameness, is merely an increased sensibility of the sole at the seat of the corn. There is no red patch of effusion, for this necessarily depends upon actual rupture of some vessels of the sensitive sole, consequent upon a severe contusion. As the smith pares the sole still more, the reddened patch may be found to extend completely into the quick; or, on the other hand, it may be merely superficial. In the former case the bruise is of recent origin, while in the latter it is of older standing.

We mentioned, in treating of inflammation, that serous fluid is poured out of the little blood-vessels of the affected part. This is the case when the sensitive sole is inflamed. A yellowish fluid oozes through the corn, and moistens the horny sole around. Sometimes so severe are the inflammatory changes, that "pus" or "matter" is formed. This is a serious condition, for, if not discovered, the pus may force its way gradually upwards to the coronet, and produce a quittor, a grave affection of the foot, of which we shall shortly have to speak. Sometimes the imprisoned matter, instead of passing upwards, may lead to inflammation of the intimate parts of the foot around it, and give rise to a very grave condition.

In cases of corns, it is first necessary to remove the shoes, and have the sole at the heel well pared away. Our treatment will now vary with the state we find the corn to be in. If matter is imprisoned, it must be let out, and the foot should be poulticed with bran for several days. A little tow soaked in tar, or in strong lotion of carbolic acid (1 in 15 of water) may be placed in the wound. When the internal structures of the foot, such as the pedal bone, are in a state of decay, it will be necessary for the veterinarian to remove the diseased tissue. In such instances, usually manifested by the discharge of fetid matter, the process of cure will necessarily be tardy and difficult, as it requires considerable professional skill, and an accurate acquaintance with the minute anatomy of the foot. When the nature of the corn has been thoroughly investigated, the animal may be shod with a three-quarter shoe. The first principle of the cure of corns is rational shoeing. After an animal has once been affected with corns, care should be taken not to press upon the particular spots, where they are alone liable to be seated.

In order to prevent corns, stoppings should be discarded, and the mistakes we have indicated in shoeing should be avoided. The web of the shoe at the heels should be broadened, and the bars should not be pared down by the blacksmith. When corns are very ubject to recur, we usually recommend the three-quarter shoe ; but, in most instances, it will be found that a plate of leather between the shoe and the sole will act as efficiently. In those cases where quittors result from corns, they must be treated in the way we shall shortly indicate.

SEEDY TOE.

By this term we understand the secretion of diseased horn, leading to the formation of a cavity within the wall of the hoof, and extending upwards to the coronet. It is called seedy toe, from the fact that it is usually most manifest at the toe, though it may extend around the whole wall of the foot. It often invades the quarters of the foot. It in reality consists in a detachment of the crust from the sensitive laminae. Seedy toe is often the sequel to laminitis, but it also sometimes follows the use of the toe clip. The sensitive laminae in this disease instead of forming healthy horn, secrete a dry soft caseous substance, which, formed rapidly and imperfectly, leads to their separation from the horny crust of the foot. A space is therefore formed, and this can readily be diagnosed by tapping the horn, when a hollow resonant note will be emitted. Professor Axe believes seedy toe to be due to the presence of small worms, to which the perverted condition of the horn is attributed. Some observers believe these worms to be accidental, and do not regard them as primary agents in the production of this disease.

Sometimes seedy toe is without doubt due to bad shoeing, by which the weight-bearing surface of the foot is limited to the wall. The progression of the animal is often but not invariably affected by seedy toe, which, it may be remarked, constitutes unsoundness. Sometimes lameness is very marked, and this is especially liable to be the case when the cavity becomes distended with accumulation of mud and sand. Seedy toe is not difficult to diagnose, as it is generally quite apparent as soon as the smith removes the shoe. The emission of the hollow sound when the foot is struck, will indicate the extent of the cavity. In these cases the diseased horn should be *carefully and thoroughly removed*, and tow moistened with a preparation of carbolic acid, or with ointment of salicylic acid, may be passed into the cavity. The foot should be kept moist by the application of some hoof ointment, and the coronet should be mildly blistered with equal parts of lard and ointment of red iodide of mercury. The animal may be shod with bar shoes. In those cases which follow founder, the hope of recovery is not so great as in others. In most instances, however, seedy toe is easily dealt with and cured. Toe-clips should be discarded.

QUITTOR.

By the term quittor we understand the presence of a diseased channel, opening upon the quarters or heels of the coronet, and extending down between the walls of the hoof, and the sensitive structures which secrete it. Sometimes the channel has but one course, while at other times it has several ramifications. It is to be borne in mind that the quittor does not open at first at the coronet, but appears as a small tumour there, which gradually comes to a point and bursts. Quittors are caused by treads ; pricks in shoeing ; corns ending in the formation of matter, which cannot escape in any other manner, than by passing upwards to the coronet ; or indeed by any injury, which ends in matter being formed either in the structures of the coronet, or in those within the hoof. Frost-bite of the coronet has also been known to lead to quittor. It will thus be seen that quittor may commence above, at the coronet, or it may commence below, and spread upwards.

A quittor is recognised by the presence of a hard, hot, and tender swelling upon the coronet. On the swelling there are soon seen one or more openings, from which is discharged matter of varying consistency, sometimes thin, and sometimes thick. If these openings are traced, it will be proved that they extend downwards, sometimes to the bottom of the foot. Quittors are distinguished from wounds or abscesses, by the presence of the little openings, which discharge an unhealthy matter ; and also by the hardness of the tumour. Lameness is sometimes extreme in cases of quittor, and the animals affected with it, which are principally heavy cart-horses, often cannot place the foot to the ground.

The treatment of quittor, which varies with the nature of the cause, requires patience and skill, as a cure is not often made before the lapse of about ten weeks or so, and may be a much longer affair, if the disease has already been of some duration. In the first place, the shoe should be removed at once and the sole pared and examined, in order to discover any possible wound, prick, or corn. If matter be found in the foot, as the result of any of these causes, an opening must be made at the sole, in order to liberate it, and allow of its escape when renewed. Then the foot should be poulticed for several days, the bran being prevented from entering the wound, by placing a piece of cloth over it. The coronet may be blistered with advantage, by means of the ointment of the red iodide of mercury. Shoeing with a bar shoe is ordered by us when the foot is much injured, and the animal seems to require it. Into the wound, it is a good practice to inject a solution of bichloride of mercury from above— half a drachm to the ounce of water, with a few drops of hydrochloric acid added. This preparation is a safe and efficient method of removing the diseased walls of the purulent channel, but it must not be repeated more than twice. If the tumour at the coronet have no opening, it will be best to make an orifice with a knife, prior to blistering the elevated and swollen tissues. When the disease does not take its origin from below, or when no prick or corn can be discovered, it is our custom to probe the wound at the coronet with the view of ascertaining its extent. The veterinarian then passes a bistoury with a hidden knife *(bistouri caché)* into the sinus, and, as he withdraws it, the instrument cuts through the diseased tissues. In addition to these measures, we may inject a solution of bichloride of mercury of the same strength as mentioned into the wound, not repeating it again unless necessary. If the wound still has an unhealthy appearance in four or five days' time, a second injection may be made. In the meantime, the foot should be enveloped in poultices, which should be renewed at least every day. There is no occasion to use strong astringent applications. Moreover, the practice of burning away the diseased tissue at the coronet with a red-hot iron, though sometimes a very good one, is not often necessary. In some protracted cases, when all other measures have been taken, and still the sinuses will not heal, it is customary to push a pointed red-hot iron to their bottom. This operation is often attended with very good results, but must be very carefully and judiciously performed. In those quittors, in which the pedal bone or the lateral cartilages have become involved, the disease is consequently of a very grave nature, and it will be necessary for the veterinarian to remove the altered structure. This serious operation is fortunately one not frequently called for.

TREAD.

By the term "tread," we understand the infliction of a wound, caused by the shoe of either fore or hind foot, upon the coronet of the opposing fore or hind foot respectively. Tread is not a common occurrence, except in heavy draught horses, and, when it does occur, it is usually traceable to overwork, or to injudicious shoeing. A severe tread, if neglected, may end in a quittor, and should therefore receive careful attention. If very slight, tread will require no treatment beyond the application of a little carbolised oil (1 in 40), or of tincture of myrrh. If the injury be of a more serious nature, the wound should be carefully cleaned with tepid water, and afterwards dresssed with carbolised oil or carbolic acid lotion (1 in 30). In those cases where "pus" tends to form, the wound must be kept very clean, and the foot poulticed for three or four days. Sometimes a little mild blistering ointment around the wound will stimulate the part to healthy action, when the healing process is unduly protracted.

OVER-REACH.

AN over-reach is a wound upon the coronet of the fore foot, caused by treading on the inner or outer edge of the toe of the hind foot. The injured spot is generally situated immediately above the heels, and is often to be attributed to careless riding or hunting over heavy country. In horses having a tendency to over-reaching, the toes of the hind shoes should be of a square pattern, with side clips if necessary. In order to prevent the infliction of this injury, circular india-rubber guards are made, which pass over the foot, and protect the seat of injury. In most instances, it will be unnecessary to poultice the foot for over-reach ; but in severe cuts, this should be done for several days. In simple cases, the wounds should be cleansed thoroughly, and afterwards dressed daily by the application of a little carbolised oil (1 in 40).

In most instances of over-reach under our notice, the animals were hunters or thoroughbreds. In some instances the tendons at the back of the heel are bruised or cut, and in such cases it will be well to rest the animal, and apply a high-heeled shoe if necessary. In these cases the wound should be well cleaned with tepid water, and carefully dressed with carbolised oil and then bandaged. If there be no actual wounds, cooling lotions (spirit 1 part, solution of subacetate of lead 1 part, water 8 parts), bandaging, and rest will suffice.

VILLITIS, OR INFLAMMATION OF THE CORONARY BAND.

THERE are two diseases of the coronet to which we must allude, before considering the nature and treatment of horn tumours. Villitis, or inflammation of the coronary band, a disease generally met with in heavy cart horses, but sometimes occurring also in more highly bred animals, is manifested by a tender, hot, and swollen condition of the coronet. The horny crust of the foot becomes harsh and brittle, owing to interference with the secretory activity of this coronary band, which is tender on pressure. The progression becomes shuffling, if both fore feet are affected. The heels are put to the ground first, but not so markedly as in laminitis. From the latter disease, villitis may be distinguished by the swollen condition of the coronet, and the harsh, dry, and striped condition of the horny crust. In these cases, which it may be mentioned are generally due to work on hard stony ground, the shoe should be removed, and bar shoes applied. Rest is essential. A mild aperient such as three or four drachms of aloes should be administered, and the diet should consist for three days of bran mashes and warm water. During this time, poultices should be assiduously applied ; but afterwards cold applications to the coronet may take their place for a time. When the inflammation has subsided, a mild blistering ointment made of three parts of lard, and one of ointment of red iodide of mercury may be applied around the coronet.

CARBUNCLE OF THE CORONARY BAND.

THE second disease of the coronet of which we may say a few words is a very rare one. It is termed carbuncle of the coronary band, and has fortunately only come under our notice on two occasions. These cases are always of great danger. It is our practice to administer a fair dose of aloes in the first instance, and to remove the shoe in order to make a careful examination of the foot. Internally, drenches composed of two drachms of carbonate of ammonium, and half an ounce to an ounce of tincture of opium may be given three or four times daily. The animal should be fed on oatmeal and linseed gruel, or indeed with almost anything he will take, to keep his strength up. Locally, the sloughing ulcers should be powdered well over with pulverised nitrate of lead, or equal parts of iodoform and calomel. Professor Williams recommends nitrate of silver. Above all things the stable should be thoroughly cleansed, and all the hygienic conditions attended to.

HORN TUMOURS.

WE have now to speak of horn tumours or keratomata, which are formations situated at the inner side of the horn of the hoof of the toe, and caused by pressure of the toe clips, or by blows. These horn tumours, usually seen at the toe of the hind feet, but not uncommonly met with in the fore ones, in many instances cause unmistakeable lameness. They press on the pedal or coffin bone, and thus cause a corresponding gap in its substance. We have not encountered many of these formations of late ; for it has been found quite possible to discard the use of toe clips altogether as unnecessary. Moreover, when still used, they are often made of less size, and are not hammered down with the violence which smiths were wont to deem it their duty to employ. Horn tumours constitute unsoundness, as, unless the coffin bone becomes absorbed, in correspondence with the growth of the tumour, lameness is manifested.

Sometimes these horn tumours are met with as the result of blows or continued local pressure, where no clips have been used. With regard to the diagnosis of their presence, which is sometimes not patent at first sight, it may be mentioned that the outer side of the hoof, corresponding to the site of the tumour internally, is often seen to be more hollow.

It will be necessary in cases of tumour in this position to remove the cause, when that is possible. In those cases, however, where lameness continues to be manifested, still more vigorous measures must be taken. The tumour may be excavated from below with the searcher from the sole of the foot, and may then be filled with tow, saturated with some antiseptic preparation. The shoes may then be re-applied, care being taken that they are adjusted, so as to cause no undue pressure at the seat of disease. When no lameness is manifested, the disease often escapes detection. We have frequently met with it in examining feet after death, when no disease of the foot was previously suspected.

PRICKS AND INJURIES OF THE FOOT.

PRICKS in the foot are of very common occurrence in horses. They are caused by nails driven into the sensitive parts of the foot, generally through the carelessness of blacksmiths. Not uncommonly, also, horses tread accidentally upon nails, or other sharp implements lying about on the ground. The writer could describe hundreds of such cases of pricks in the foot, which have come under his care ; but it will suffice here to speak of the subject in a general way, indicating at the same time the method of treatment to be adopted. It should be remembered that injuries of the frog or sole of the foot very frequently cause extreme pain and lameness, and must never be neglected ; for, apart from all risk of lock-jaw setting in, very serious constitutional disturbance and rapid increase of the local mischief, are apt to follow in neglected cases.

Although pain and lameness often follow immediately after the infliction of the injury, they may not become manifest for several days afterwards. Local inflammation is set up in the region of the prick, and then "matter," technically known as "pus," is formed. This, being imprisoned by the horn, causes intense pain by the pressure it exerts on the surrounding parts. Sometimes a horse is pricked, and the smith perceiving it at once, draws out the nail, while at other times the nail is left in. In either case, whether the nail be left in or not, more or less inflammation is of necessity set up. Again, at other times a nail when driven into the horn splits, and while one arm passes in the proper direction, the other passes into the sensitive parts, and likewise sets up inflammation. Necessarily the signs and results of a prick will vary exceedingly, not only according to the seat of injury but also to its depth. The writer has seen a number of instances, where the njury and its results were confined to a very small area. In neglected cases, matter may be developed under the whole of the sole of the foot. As a rule, a prick is not difficult of detection, though it may be mentioned that sometimes, after being shod, horses may go lame, when the heels have been very much pared down, although there be no prick or injury whatever.

Before mentioning the usual signs of a prick, we may shortly consider some of the risks encountered in nailing on the shoe according to the English method, briefly comparing it with the Arabian plan. "In warm countries," writes Mayhew, "the horse's hoof grows strong and thick, and the wall is allowed to descend half an inch below the sole. Completely through the portion of the projecting hoof, the untutored Arab drives the nails to secure the shoe. Proceeding thus, he does not injure the foot by the insertion of foreign bodies through its more brittle substance, while he secures both the resistance and tough qualities of the complex covering of the foot. The English smith, on the contrary, by ranging the holes for the fastenings round the edge of the shoe, drives the nails into the harder kind of horn and transfixes the crust for a considerable distance. The English shoeing nail is intended only to pierce through the black or outer substance of the wall. Now, though this may seemingly afford the better hold, it also offers the more dangerous dependence." There is, moreover, the risk of pricking the sensitive parts, when the nail happens to turn a little to one side, as well as of driving it "too fine;" that is forcing it too near the white horn, rather than of directing it through the centre of the narrow dark crust. The smith ought, in shoeing a hoof with thin walls, to exercise the greatest care not to injure the sensitive parts by pricking or by driving the nail "too fine;" for a nail when driven "too fine" may bulge inwards, when the animal is worked, and inflammation then setting in, severe lameness and the formation of matter are sometimes induced.

Mr. Mayhew, did not advocate the Arabian method, but he pointed out that the drawing knife might be used with more caution, and he saw no reason why the wall need be cut away until level with the horny sole. The latter, by being thus exposed close to the earth, is frequently injured. He suggested on these grounds that half-an-inch of crust should be allowed to project below the sole, which should be of moderate thickness. The idea that the breadth of the shoe affords the slightest protection should be at once abolished, and the shoes should be made just wide enough to afford protection to the wall. With these rational views we entirely coincide.

We may point out, in respect to levelling both sides of the lower surface of the hoof, that the difference of a few fractions of an inch between them may lead to very untoward results. A blacksmith should always be careful to ascertain whether the foot is level or not, because undue strain is imposed on the joints and ligaments when there is unequal pressure, and, moreover, the hoof tends to become deformed, and the growth of the horn modified (Fleming).

In the following plan, to show how the hoof should be levelled at each side, in order to preserve the proper direction of the limb and foot, the line A A is seen to be at right angles to the vertical line B.

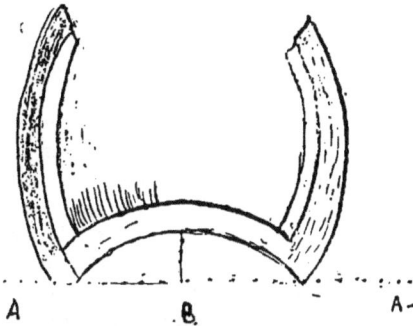

A   B.   A-

We may now proceed to speak of the usual signs and methods of detection and of treatment of pricks in the sole and frog. Very often an animal, as soon as he is pricked, flinches and goes lame from the pain inflicted ; and the nail when withdrawn is sometimes blood-stained, showing that it has taken a wrong direction. When an animal goes lame after being shod, we may frequently find the offending nail, by tapping lightly with the hammer round the hoof ; and we may endeavour to define the seat of the injury more exactly, by pinching the crust with the pincers, in the region of the suspected spot. When the shoe has been removed, "matter" not uncommonly oozes from the hole made by the intruding nail ; but the "matter" will of course not yet be formed, if the injury be of recent standing. When the injured spot is found, it will be necessary to pare out the puncture with the searcher, at the same time being very careful not to injure the sensitive parts. By this means the "matter" is liberated. If it is still left pent up in the foot, quittor, and still more extensive disease of the structures within the hoof, will most probably ensue. Some practitioners prefer to cut down upon the nail from the outside of the hoof, with the view of running less risk of injuring the sensitive parts of the foot. We, however, do not recommend this method of procedure, but prefer the usual method of cutting away the separated horn. If the injury be not serious, and there be but a little "matter" oozing out of the hole, but not very great lameness, a little tow, saturated with tincture of myrrh, may be passed into the wound, and the foot carefully poulticed with bran. Sometimes blacksmiths and others use turpentine, or certain very deleterious mixtures for dressing such wounds. *We are now attending a case of severe lock-jaw in a yearling thoroughbred, valued at £1,000. The owner had been dressing an injury in the sole with turpentine, for a fortnight before the disease manifested itself. We cannot too strongly condemn such practices, as we have repeatedly seen the injured member made ten times worse than before, by such ill-devised means.

In severe cases, the above-mentioned simple methods of treatment are of course not applicable ; though in all instances it is necessary to pare out the injured part, and poultice the foot, until the inflammation and lameness

---

*We are delighted to be able to record that the foal referred to made a complete recovery.

M

subside. In cases of injury of the foot, it is well to give a moderate dose of aloes in the first instance, and to feed the animal on a laxative diet of warm water, bran mashes, and oil-cake gruel, until the inflammation and fever subside. Half an ounce of nitre and half an ounce of bicarbonate of potassium, may be given once daily in the drinking water. The writer was recently called to a case where it was necessary to remove the whole of the sole, and the animal, although previously much neglected, made a complete recovery. In some instances, the coffin bone is injured by the penetrating nail, or other foreign substance. Such cases, as a rule, are very severe and lingering. In a horse recently attended, a nail had penetrated into the navicular joint, and caused not only very acute pain, but also very high fever. In such cases, even when the "matter" has been liberated, the animal still goes very lame ; and, indeed, the continuance of thelameness is sometimes the only symptom, which leads us to suspect such a serious condition of the foot. During the early part of last year, we were called to see a six-year-old cart mare. A piece of pointed wood had penetrated into one of the feet, between the bar and the side of the frog, for a distance of about three inches. The pulse was imperceptible, and the mare gasped for breath in her intense agony. In a very short space of time, in spite of all remedial measures, the animal died from the acuteness of the pain. Although strongly recommended by me to shoot the animal, the owner had refused, not realising the futility of treatment. Some years ago the late Mr. D. Gresswell was called to a horse with acute lock-jaw, the result of a nail which had passed into the cleft of the frog. We may conclude our remarks on injuries of the foot, by advising our readers in all severe cases, to procure professional aid as early as possible.

SIDE-BONE.

WE mentioned, in describing the structures of the horse's foot, that the pedal or coffin bone, contained within the hoof, has, on each side of it, a lateral prolongation of cartilage or gristle. We may now add a few particulars regarding these important appendages, which are generally spoken of as the lateral cartilages. These are thicker and more extensive in the fore than in the hind feet, and are peculiar to the equidæ or horse tribe. When one considers the important purpose which these cartilages subserve, it will readily be seen how it is that, if they are ossified, or, in other words, turned into bone through disease, when they are called " side-bones," very untoward results are produced. Regarding the functions of these two thin quadrangular plates of cartilage, which surround the wings of the pedal bone, Professor Williams says, that, in virtue of their elasticity, they assist the sensitive frog and the soft structures of the foot, in regaining their natural position, after being pressed upwards and outwards, by the weight of the animal. Undoubtedly, he writes, they expand at their hinder borders, each time the animal puts his foot to the ground ; but, in this expansion of the heel, they are mere passive agents, being in fact pressed outwards by the structures, contained in the space between them. They are, however, active agents in causing the contraction of the heel; for, when the pressure is removed from their inner surfaces, they tend to assume their natural position, in virtue of their elasticity, and the pressure they exert upon the sensitive frog, forces the heel into its original shape.

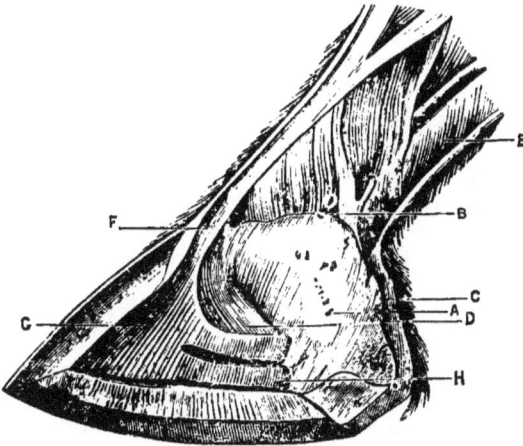

CARTILAGINOUS APPARATUS OF THE HORSE'S FOOT.

A, external face of the lateral cartilage; B, superior border; C, posterior border; D, anterior lateral ligament bordering the cartilage in front; E, flexor tendons; F, extensor tendons; G, coffin-bone.

Briefly, then, they may be said to expand, when the foot is on the ground; and to assist contraction, when the weight which forces the sensitive frog upwards and outwards, is removed from the foot. Professor Williams, in short, holds that these lateral appendages act, as it were, as "elastic sides," preventing undue expansion of the soft parts of the coronet and heel.

The term side-bone, we have said, denotes a bony or ossified condition of the lateral cartilages. This condition is commonly met with in heavy draught horses, and is but rarely seen in the lighter breeds. It is almost always met with in the fore feet, though in rare instances it has been observed on the hind ones. In the latter situation, it is never known to occasion lameness. The lateral cartilages are of lesser size here, and, being of less functional importance, are consequently much less liable to become diseased.

We may now proceed in the first place to examine the causes of this very common form of disease among our heavy draught horses. Some authorities compute that over fifty per cent. of the heavy draught horses become affected with this disease by the time they have attained the age of six or seven years; but, according to our own computation, sixty per cent. is not an exaggerated estimate of this common form of morbid action. Why is this? Indubitably this morbid process depends, as do many of the other diseases of the foot, of which we have already spoken, on the violent and continued concussion on the hard roads. The heavy weight of the animal, and the shoeing with high heels or calkins, are additional factors in the causation. High calkins deprive the foot of the uses which the frog serves as a buffer, and the concussion, received at every step by the heels, is thus directly transmitted to the cartilages, which suffer in consequence. The pressure on the heels is, moreover, greater than it would otherwise be, were high calkins dispensed with. Again, the sensitive frog is pressed downwards, by this practice of using high calkins, and the horny covering, being elevated from the ground, does not afford the support it otherwise would do. As in so many other diseases of man and animals, hereditary influence also, no doubt, predisposes very strongly to the contraction of this form of bony degeneration. The practical conclusion to be drawn from this fact is, that one should not breed from animals so affected.

The formation of a side-bone is often spread over a long period of time. When met with in aged cart-horses, whose progression is often thereby not much affected, they are not of any great moment. When, however, they are met with in the lighter breeds of horses, whether they cause lameness or not, and when they affect the gait of the cart horse, they are of much more importance. Now, although side-bone constitutes unsoundness, it is not necessary, or even advisable, to condemn an animal as unsound, unless the progression be affected thereby. Side-bone is in most instances accompanied

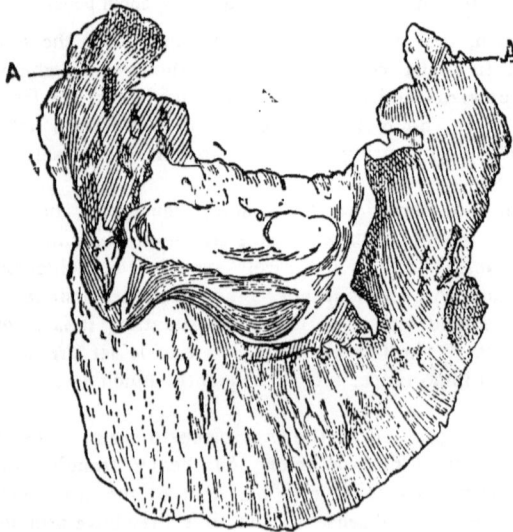

Pedal bone of the horse, showing the ossification of the lateral cartilages, A, A.

by lameness in harness and in saddle horses; though when they are not worked, there being no concussion, the progression is not necessarily impeded, or altered. Mr. Fearnley writing on these points says, "we not unfrequently find the lateral cartilages strong, but yielding, and, when that is the case, a horse with a good foot otherwise may be considered as sound. These strong lateral cartilages are not ossified, and have no particular tendency to become so. If you can feel them to yield, no matter how little, they are not ossified." Very different, however, is it with heavy-bodied dray horses, in which the lateral cartilages have a strong tendency to become transformed into bone.

A harness or saddle horse, although sometimes not actually lame from side-bone, will generally lose his elasticity of action when worked; and, before long, actual lameness is to be expected, if not already manifest. In a cart-horse employed for slow work, it is not of such paramount importance that the action be characterised by that elasticity natural to the healthy foot ; but, if the soles be flat or convex or otherwise misshapen, and the action of the animal be stiff, he cannot be passed as sound.

In examining a horse for side-bone, the lateral cartilages should be pressed upon firmly. If normal, they will be found to be yielding and elastic. In disease they become hard and inelastic, owing to the deposition of bone ; and a hard swelling may be found at the back of the coronet and heels. If the morbid process be recent, and in a state of inflammation, the swelling will be found to be tender and hot. It must be pointed out that sometimes the bony deposit involves the whole cartilage uniformly, while at other times it affects only one or more isolated parts of it. Again, sometimes it involves the hind portions ; and at others it only affects the fore parts of the cartilage, in which case the hardness is felt at a point well forward on the quarter. In the latter position, side-bone is much more likely to cause lameness, than when situated more posteriorly, and in this situation has sometimes been mistaken for ring-bone, a disease on which we shall shortly speak. These two affections, side-bone and ring-bone, however, are entirely different, involving different parts, and occasioning different kinds of lameness. An animal, when lame as the result of side-bone, brings the toe of the foot first into contact with the ground. When both feet are affected, the action resembles that of navicular disease, each of these diseases been characterised by a want of elasticity of action, and by a short groggy style of progression. Sometimes, it may be added, side-bone affects only one lateral cartilage.

The animal should be shod with bar shoes, and be rested. The affected part should be smartly blistered with ointment of biniodide of mercury. If these measures are not effectual, firing will be necessary. Prick-firing, or firing with a small pointed instrument, is the method generally best adapted for the cure of this affection. When thoroughly applied, this is found to be, in many instances, a very efficient method of treatment. Sometimes other structures are involved in the disease of the cartilage, and in these cases it is better to employ stripe-firing.

In hopeless cases, neurotomy may be performed as a last resource. It has been found that this operation is often more successful, in the relief of lameness from side-bones, than when the result of navicular disease. The French veterinary surgeons sometimes excise diseased lateral cartilages ; but this operation is not one of much practical value.

RING-BONE.

By ring-bone we understand a bony or osseous deposit of an inflammatory origin, formed upon the upper and lower pastern bones. This disease generally affects the hind pasterns, but may be found on the fore ones. There are two kinds of ring-bone, named "true" and "false" respectively. By a false ring-bone, we mean a bony growth which is developed on one or both of the ridges situated at the back of the long pastern bone. This form of ring-bone does not always cause lameness ; but, when large, not uncommonly affects the progression of the animal. Although, according to Mr. Fearnley, "ring-bone is an unsoundness which cannot for a moment be regarded in any mitigated light ;" and, although wherever situated, it very commonly affects progression by impeding the action of the ligaments of the joints or of the tendons, it is, nevertheless, when of the false kind, not invariably to be regarded as an unsoundness.

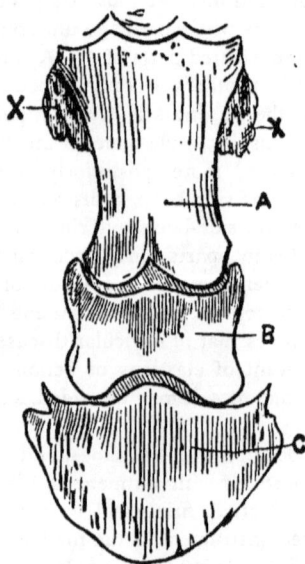

In the figure A is the long pastern bone, B is the short one, and C is the pedal bone ; X shows a false ring-bone.

True ring-bone is the term applied to a deposit of bone in either of two situations. When the deposit of bone involves the pastern joint, that is the joint between the two pastern bones, it is termed high ring-bone. This is the variety most commonly met with.

When the deposit affects the coffin-bone joint, that is the joint between the small pastern and the coffin-bone, it is termed low ring-bone.

Although this latter form is necessarily the more grave variety of ring-bone, yet it must be remembered that true ring-bone in either situation always constitutes unsoundness, as it occasions very inveterate and often incurable lameness. In some instances, both high and low ring-bone coexist at the same time.

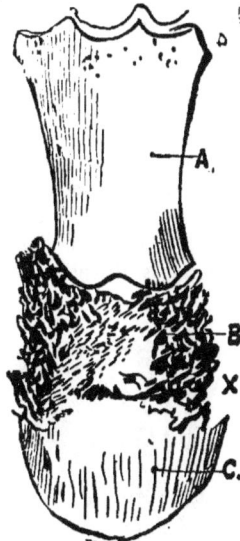

In this figure X shows the position of low ring-bone.

Speaking of the nature of side-bone, we showed that it is a disease of the side cartilages of the coffin-bone ; and it will therefore be seen that it is of a totally different character from true ring-bone, which [is] a bony deposit around the ends of the bones forming the pastern and coffin joint. When the deposit involves the latter joint, which our readers will remember is within the upper part of the hoof, the lameness is often very severe ; because the horn, although elastic, nevertheless, fitting closely, presses upon the new growing bone.

Ring-bones vary greatly in size and shape. They are generally confined to the sides and front of the bones ; but sometimes they extend to the back of the joints, forming a complete "ring." Hence the name has been derived. Sometimes only the sides of the bones are affected, and sometimes only the front parts of the joints are invaded by the bony growth. When the front part of the bone is affected, the lameness is necessarily very severe. Yet it must not be thought that the degree of lameness depends upon the size of the bony matter thrown out. We have known cases in which a large deposit gave rise to little or no lameness, and many cases where but little new bone caused very severe lameness. In cases where the lateral parts of the bones are only affected, lameness is often not so marked as when the ring is complete, or when the deposit is only formed on the front of the joint. It must be borne in mind that ring-bone is the result of inflammation, affecting the ends of the bones. Lameness is therefore manifested at an early date, before any bony enlargement can be felt, as the result of the inflamed condition of the bone. The progression will remain affected, as the bony growth continues to be formed and deposited; but, when this is completed, and the joint has become fixed and immovable, the action may be but little impeded. Indeed, the lameness in some instances disappears altogether, although the gait is not as elastic as it was before. We have said above that side-bone is in most cases found on the fore legs; and in this it differs from ring-bone which is somewhat more common on the hind than on the fore feet. When a horse is lame from a ring-bone in the fore extremity, he invariably goes on his heel, excepting in those instances in which the deposit is at the back part of the bones. When the hind limb is affected, the animal brings his toes down first, when the pastern joint is involved, and the deposit does not involve the front part of it; but, when the coffin-joint is diseased, the heel is brought to the ground before the toe.

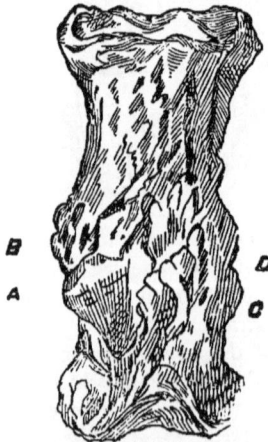

The figure from Percivall shows the back of the pastern joint, affected with ring-bone, A, B, C, D.

Regarding the causes of ring-bone, we have not much to say. Hereditary influence, however, it has been proved, is a potent agency as a predisposing factor. As in the case of side-bone, therefore, the practical conclusion to be drawn from the fact, is, that one should not breed from animals so affected, unless the disease be traceable to some actual injury, inflicted by accident.

In addition to heredity as a factor in the causation of this disease, it has been noticed that horses with straight upright pastern bones are more likely to contract ring-bone.

In cases of ring-bone, it is often impossible to do very much to alleviate the lameness ; but it is advisable to fire deeply in the first instance, and then blister smartly with ointment of biniodide of mercury, with the object of promoting the absorption of the deposit, or of causing cessation of the inflammation.

Neurotomy has been recommended for chronic cases of ring-bone ; but we have not much faith in its value, although in some instances it is said to have proved successful.

CHAPTER III.

WOUNDS.

General remarks on the Treatment of Wounds. Sutures, Antiseptic Applications. Brushing and Speedy Cutting, Sore Back, Sitfast, Harness Galls. Broken Knees.

GENERAL REMARKS ON THE TREATMENT OF WOUNDS, SUTURES, ANTISEPTIC APPLICATIONS.

THE subject to which we now call the attention of our readers, is one of universal interest. All horsemen should have some accurate knowledge of the usual scientific methods of treating the commoner and less severe kinds of wounds.

With the object of being more precise in our description of wounds, we may conveniently divide them as follows :—incised, or made with a cutting instrument, punctured, lacerated, bruised, and finally, those caused by firearms. We might also add poisoned wounds to this list.

A minute description of the ways in which wounds are healed, would doubtless be of great interest to some of our readers ; but, as we fear this would not prove of much practical value, we shall forthwith proceed to consider the best methods to be adopted for promoting the repair of the injured tissues. It is well known, that the power of repairing lost tissues and the healing of wounds, is much greater and more rapid in some of the lower than it is in the higher animals, such as the horse and ox. If a crab or lobster have the misfortune to lose a limb, this can again be reproduced ; whereas, as we ascend the scale of animal life, the faculty of restoring a lost member gradually disappears, and is finally lost altogether. Nevertheless, the healing of injuries of a very severe and extensive kind, is of daily occurrence in horses and other animals. Our methods of treatment of to-day are in accordance with the dictates of practical science, and more especially with those discoveries which, intimately associated with the name of Sir Joseph Lister, have shown the supreme value of great cleanliness and antiseptic applications. Our forefathers, unfortunately, had no knowledge of those tiny little fungi spoken of as germs. Certain organisms, floating about in the air around us, find their way into wounds, and thrive and ferment the more, as they find the raw surfaces unclean and unhealthy, and then may enter the blood vessels of the animal, where they may multiply, and cause great constitutional disturbance, and even death. In healthy wounds, fortunately, they cannot thrive. Hence we see the value of maintaining cleanliness, and of applying antiseptic lotions, to prevent their becoming established, and increasing rapidly.

When our attention is called to the existence of a recent wound, we may find it bleeding, or the blood may have already ceased to flow. In most instances, moderate pressure for a time will stay the bleeding, or the application of a mixture of tincture of perchloride of iron one part, and of

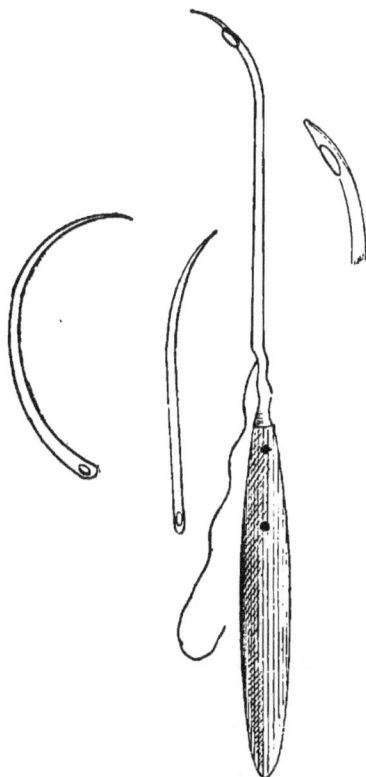

The above illustrations show, first, two common suture needles, and secondly, Simpson's needle.

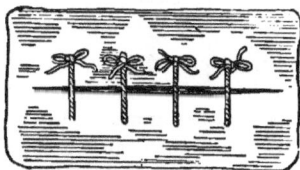

Interrupted Suture.

water six parts, will act as a powerful styptic in arresting hæmorrhage. Pressure, we may mention, is more often employed when the wound has been sustained on one of the limbs. When blood spurts from a wound in jets of a bright red hue, an artery is injured, and in order to stay the hæmorrhage, it must be tied. When an artery is cut in two, blood, as a rule, does not escape in jets, because the divided ends contract, in virtue of their elasticity, and moreover they become retracted also, inasmuch as the vessel is in a permanent state of tension or stretching. Our readers will thus see that it is when an artery is partially divided, that hæmorrhage is liable to be so severe and continuous. Sometimes it will be very difficult to find the bleeding artery, and in such cases the application of the red-hot iron may arrest further hæmorrhage. After docking, this method of closing the divided arteries by searing is commonly adopted. In the next place, it is advisable to sponge gently over the wound with tepid water, in this way removing any dirt or blood clots, which may remain in the injured part.

The steps now to be taken will vary much with the nature and extent of the wound. If it be incised, our object will be to bring the several parts together; and this may be accomplished by sutures or bandages, or by plasters in trivial cases. If the wound be very deep, it is customary not to sew up the severed tissues for several hours, in order to allow time for the escape of the liquid serum, which oozes from the injured parts. Strips of plaster are especially adapted for bringing together the edges of a wound, when of a very superficial nature. When it is necessary to employ sutures, we may use what is termed the interrupted, the twisted, or the continuous method.

The twisted sutures our readers will observe, on referring to the pictures below, is made by inserting a curved pin through the lips of the wound brought together, and then maintaining its position there, by winding thread between the two ends in the form of a figure 8.

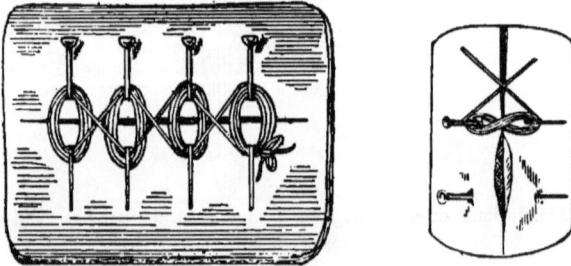

Twisted Suture.

Sutures are not so much employed in veterinary as in human practice, as it is difficult in many instances to maintain complete rest, when the injury is seated in some parts of the animal. When the eyelid is torn, as it often is, or when the nostril is rent open, and in many injuries of a like kind, it is of course absolutely necessary to stitch up the severed tissues as early

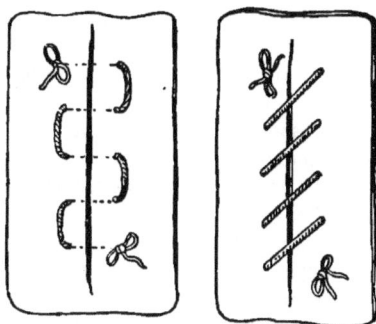

Continuous Sutures.

as possible, first, however, carefully sponging the raw surfaces with tepid water. We have had several cases in foals, where very large rents extending from the edge of the mouth to the middle of the cheek, needed to be carefully sutured together. In one instance, owing to the motion of the cheeks in mastication, the wounded surfaces had to be again sewn up, as the sutures all became loosened in a few days ; and in another case three successive suturings by the interrupted method, were required at intervals of about a week, before the tissues grew firmly together.

For suturing we often use medicated strong twine or silk, but in some instances silver wire is to be preferred. In these operations, one should commence the stitch about half an inch to an inch, varying with the thickness of the lips of the wound, from each edge, and should not be afraid of passing the needle pretty deeply, so as to obtain a sufficient hold. As a rule, the stitches should be about half an inch to an inch or so from each other ; and one should be careful to bring the corresponding parts of the severed tissues into close apposition with one another. When the sutures have been carefully made, we may bathe the tissues with a bland unirritating antiseptic lotion, or may anoint the part with a little ointment. A lotion of boric acid is very useful. This may be made of boric acid, one part ; hot water, twenty parts. Dissolve, and when cold, use the clear solution. A lotion of boroglyceride, made of one part of this preparation with thirty parts of water, is likewise very efficient. A very useful ointment of boric acid may be made of six parts of vaseline to one of the acid. After dressing the wound, it may, if necessary, be carefully bandaged; but this will seldom be requisite. No fomentation should on any account be applied, so long as the wound remains free from inflammatory action ; but the surfaces may be bathed with the lotion once daily, or more frequently. The sutures may be removed in about eight to twelve days after being inserted.

Having now disposed of the different methods of suturing, let us turn to consider more closely some points regarding the antiseptic treatment of wounds. We have here especially recommended lotions and ointment of boric acid in preference to carbolic acid, because they are much less irritating, when applied to recent wounds. Nevertheless, carbolic acid lotions and oils are very valuable, more especially when the injury is taking on an unhealthy action, or is discharging fetid matter. A useful lotion of carbolic acid for veterinary purposes may be made of carbolic acid, one part; water, thirty-six parts; and glycerine, four parts. For superficial injuries, carbolised oil is sometimes to be preferred to the lotion, as it does not flow away so rapidly, or evaporate to the same extent as the former preparation. It may be made of olive oil, thirty parts; and carbolic acid, one part. For foul ulcerated surfaces, twenty parts of oil to one of the acid will be found a valuable application. Lotions and ointments of oil of eucalyptus or of salicylic acid are also very valuable.

In cases of incised wounds, the animal should be fed on a laxative cooling diet, and the bowels should be gently acted upon, by two or three drachms of aloes. Punctured wounds are of a more dangerous nature than simple incised injuries. In those instances where the tissues are not much lacerated, it should be our object to promote early adhesion by the application of weak boric acid lotion, and bathing with cold water. Suturing will necessarily not be applicable to such cases. In very severe punctures, the danger is much greater, and it is very important to apply warm water fomentations assiduously during the day. In case any foreign body be left in the wound, it must be removed as early as possible ; and, if there be severe hæmorrhage, steps must be taken to prevent it. Poultices and fomentations are also of value in those cases, where the injury may be expected to take on an inflammatory action. Internally, a mild aperient of three or four drachms. of aloes should be given, and strict quietude should be enjoined. If there be inflammatory action and febrile symptoms appear, the diet should be laxative and restricted in amount, and drenches, containing five minims of Fleming's tincture of aconite with four ounces of liquor ammonii acetatis, may be given with four ounces of water twice daily. When a limb is much injured, it is sometimes advisable to place the animal in slings.

In simple bruises, cold applications are indicated, such for example as spirit lotion, which may be made of spirit, one part ; solution of acetate of lead, one part ; water, eight parts. If, however, "matter" or pus is already being formed, warm applications and poultices are necessary. When the injured animal is much debilitated, strengthening diet and tonics soon become necessary, especially if there be much discharge of matter.

A very severe incised wound came under our notice just lately ; and we may conclude with a short *resumé* of this important and interesting case :—On November 17th, 1885, we were summoned to see a thorough-bred yearling foal, on a farm on the Lincolnshire Wolds. The muscles in front of the near fore leg, between the shoulder and the knee, were quite divided to the bone, and hung down pendulously about seven inches. The skin was torn transversely and longitudinally. This severe injury had been sustained several hours before our arrival. In accordance with the usual prevalent but most erroneous popular notions, the furnace had been lighted and cloths procured, for the purpose of continuous fomentation. Happily this had not been started ; but the fact of its being strongly discountenanced, occasioned great surprise to the owner and his servants. The severed muscles were bathed, and stitched together, with carbolised cat-gut sutures. A drainage tube smeared with ointment of carbolic acid, eucalyptus, iodoform and lard was inserted over the muscles ; and then the skin was stitched up with medicated silk on the whole extent, with the exception of the lower part, through which the tube was left depending, to act as a draining orifice to this extensive and severe injury. The external surface of the wound was now covered over with a bundle of carbolised tow. The wound was then carefully bandaged up, so as to support the lower part of the disunited muscles. On the following day, the injury showed no alteration beyond slight swelling. The parts were dressed with a solution made of carbolic acid, eucalyptus oil, a little tincture of opium and water, and the tube was re-dressed. On November 20th, there was a little more swelling, which had broken several of the sutures in the skin. There was now some discharge externally. No fomentations were allowed, but the parts were dressed daily with the antiseptic ointment. The foal was one of great value, and had been entered for racing ; and consequently an attendant was set aside to watch, and attend to him constantly. On November 21st, the skin sutures had all broken away, but the union of the muscles appeared to be quite firm. The pulse rose to 48 beats per minute, but the temperature always remained at its normal height. After this time, the parts were only dressed with the antiseptic ointment above mentioned. On November 24th, the union of the muscles was firmer, but the skin had separated about four-and-a-half inches. Much granulation tissue, otherwise called proud flesh, had now formed. On the 27th of November, the discharge had almost ceased, and on the 2nd of December, it had quite disappeared. The wound was healthy, and the skin wound was now only two inches long. On December 9th, the foal was liberated. Afterwards, the remaining tissues speedily grew together, and ultimately the animal made a perfect recovery.

BRUSHING, SPEEDY CUTTING, SORE BACK, HARNESS GALLS.

WE now propose to consider briefly the nature and methods of treating several forms of injury to special parts. The first kinds to which we have to draw attention are brushing and speedy cutting. Of these two unpleasant forms of self-inflicted injury, the latter is the most dangerous.

By brushing we understand the wounding of the fetlock by the outer edge of the inner quarter of the shoe of the opposing leg. This injury is chiefly confined to the hind extremity. When the animal wounds the inner side of the fore leg immediately below the knee, by the agency of the opposing fore foot, the injury is termed a speedy cut. As might be expected, horses not uncommonly inflict a wound, at a point between the seats of these two injuries. In some cases of spinal disease, cutting is liable to be very severe indeed. In such instances the injury is inflicted by the whole of the hoof, and not only by the tip of the shoe.

Brushing is frequently due to weakness, and is, therefore, especially common in long-legged, debilitated animals. When exhausted after a long journey, many horses are liable to cut, and often very seriously. Brushing may also be due to turning out the toes, or to certain irregularities in the shape of the animal. Wide-chested horses, with well proportioned hind-quarters, very seldom cut. This habit sometimes, moreover, owes its origin to defective shoeing, by which the outer quarter of the foot is made higher than the inner. After treating the injured spot by the application of some antiseptic ointment, as the unguentum acidi borici, it is necessary to take steps to prevent the infliction of this injury. A very valuable antiseptic ointment, useful for dressing the injured part, may be made of oil of eucalyptus, two drachms ; carbolic acid, half a drachm ; iodoform, half a drachm ; lard, an ounce and a half ; vaseline, an ounce and a half. If there be a large scabbed surface, caused by the infliction of previous cuts, this is removed by poultices, before applying the ointment. The formula above-mentioned, we may add in passing, is a very valuable application to any sore surface, as it possesses great antiseptic and healing properties.

With regard to the prevention of brushing and speedy cutting, it is found that horses shod by the Charlier method—of which we spoke in treating of navicular disease—seldom or never inflict these injuries upon themselves. The patent pads made of india-rubber, are very useful in preventing speedy cutting. They are shaped like crescents, and consist of two distinct parts, one flat, the other projecting in the form of a pad. The flat portion is introduced between the shoe and the foot, and the pad thus projects beyond the shoe. If the owner does not procure these valuable preventive pads, the inner side of the shoe of the injured limb may be made thicker, or the horny crust of the outer quarter of the same foot may be made lower ; and that section of the shoe which inflicts the wound must be smoothed off by the smith. In case these alterations prove unavailing, a stout india-rubber ring, such as that commonly employed, or a leathern boot laced on the leg may be procured.

Although we mentioned that in speedy cutting the injury is generally situated below the knee, we have met with instances where it has been inflicted just above the joint, and it is not so very uncommon for the hind limbs to be similarly injured, immediately below the hock joints. The injury, especially when repeated, is liable, like brushing, to cause a bony growth at the wounded spot ; and its repetition increases the tumefaction, and renders the habit more liable to become permanent. In examining horses, it not unfrequently happens that a bony growth is found at the spot, where a speedy cut has been inflicted by the shoe. It is advisable, therefore, in purchasing a horse, to see whether there be any traces of previous injuries.

In some severe instances, speedy cutting is attended with some constitutional disturbance, and cases in which matter is formed at the bruised spot are not very uncommon. In such cases the animal must be rested, and placed in a comfortable loose box. Good nutritious diet of a laxative nature, such as oil cake gruel, should be allowed, and internally some tonic medicine will often prove of great service. In serious cuts the bruised parts should be fomented with warm water ; but, if not severe, the application of the above ointment will be sufficient. Should any matter be formed in the injured part, a horizontal opening at the lower part is necessary, in order to allow it to escape. Afterwards cooling and astringent lotions are very valuable, and may judiciously be combined with steady pressure by means of bandages.

By the term "warbles" are meant swellings caused by undue localised pressure of the saddle or collar. If the irritation continue, the tumour may suppurate, and matter be formed. Thus a more serious state of things is set up, necessitating rest and careful treatment. In most instances, cooling applications and removal of the pressure will relieve the inflammatory condition of these swellings ; but, when this process is more severe, warm water fomentations and poultices are indicated. When the swelling remains unabated, and matter threatens to form, the ointment of oleate of mercury may be applied. When formed, the matter must be liberated by the knife, and poultices assiduously applied. To hasten the healing process, the antiseptic ointment above-mentioned, (viz. that of eucalyptus, iodoform, and carbolic acid) will prove useful. Afterwards, as the sore heals, astringent lotion of alum, tincture of myrrh, and cold water will harden the disordered tissues. During the process of cure, the horse must have no saddle work. By way of preventing these swellings, the saddle may have a thick piece of felt stitched to the pannel on each side. When a "warble" is neglected, it assumes a chronic unhealthy appearance, and is termed a sitfast. This unhealthy condition should be treated by poultices, until the scab be removed, when the antiseptic ointment should be applied twice daily.

A good application in the first instance for sore shoulders, is a lotion composed of glycerine one ounce, solution of acetate of lead one ounce, methylated spirit one ounce, and of water eight ounces.

BROKEN KNEES.

WE may now turn to the consideration of broken knees, unfortunately a very common form of injury. Our readers will remember that we said, in speaking of the horse's knee, that it corresponded with the human wrist, and is formed by seven small bones, arranged in two rows. The upper row has three bones with an additional one at the back, while the lower one has three independent solid osseous components. A very important tendon passes over the front of the knee, and when called into action, extends the joint, which is, properly speaking, composed of three joints. Between the tendon and the knee there are two so-called bursæ, or pockets containing lubricating oil. Broken knees, under which term we include slight as well as grave injuries to the knee, caused by a fall, or otherwise, very commonly leave, after healing, some evidence of their previous existence.

N

It is necessary to be on one's guard in purchasing a horse, to see that he has never sustained an injury of this kind. We do not necessarily consider a slight roughness of the skin over the joint, as constituting unsoundness, except when it interferes with the action of the animal. It must, at the same time, be borne in mind that a somewhat severe injury does not always leave a large scar; and the joint, therefore, may be much weakened, and the progression of the animal rendered insecure, by what appears as a very slight visible blemish. As our readers know full well, any blemish of the knee reduces the value of a horse very materially.

Lastly, we may turn our attention to the consideration of the treatment of broken knees. In the first place, the injury should be bathed with tepid water, until all the grit and dirt have been gently but thoroughly removed. The animal should then, when the injury is at all severe, be tied up, and fed on laxative food. In such instances it is advisable also to give a gentle aperient, say three drachms of aloes, and to place the animal in a cool, airy box, with the head tied up. The borax ointment, we have already mentioned, is a valuable local application. After anointing the wounded surface, a strip of lint soaked in carbolised oil (1 in 25) may be placed over the wound; over this a bandage may be gently applied, with the view of keeping the application in contact with the wound. The dressing should be repeated once daily. When the injury is very severe, slings are necessary, as the animal, becoming exhausted, may be unable to remain standing until the healing process is completed. When the joint is open, and the oil which lubricates it escapes, the injury is necessarily much more grave.

Sometimes the bones of the knee are actually broken in the fall. These cases seldom recover. The tendon on the front of the joint is sometimes much lacerated and bruised, and this also is a source of additional danger to our patient. Warm water fomentations are not to be applied to the wound, except when the joint becomes immensely swollen and inflamed. In some instances, sutures have been employed for sewing together the divided skin, when this covering is alone injured, and torn in the fall. They are, however, not much good as a rule, because, when the animal bends his knee, they usually burst, and the rent is made worse than before. With the view of stimulating the healing of the wound, if it appears sluggish, the antiseptic ointment of eucalyptus, carbolic acid, and iodoform is very useful. It also stimulates the growth of the hair afterwards. The application of caustics to the so-called proud flesh is an unnecessary performance.

CHAPTER IV.

SPRAINS.

General remarks on the Nature and Treatment of Sprains. Sprain of the Suspensory and Check Ligaments. Curb. Sprained Back. Sprain of Fetlock and Hock. Sprung Hock. Sprain of the Shoulder and Elbow Joints.

GENERAL REMARKS ON THE NATURE AND TREATMENT OF SPRAINS.

In dealing with the important subject of sprains, we shall adopt the same method as the one we followed in treating of wounds; firstly, speaking of sprains and their treatment generally, and then turning our attention to the elucidation of the nature and therapeutic measures, necessary for the treatment of these unfortunately common injuries. No doubt our readers are aware that the muscles, tendons, and ligaments are the structures which under certain circumstances may be "sprained" or "strained." The muscles are endowed with contractile power, and by means of this, the bodily movements are executed. Had we space at our disposal, we might enter shortly into a consideration of the microscopical features and physiological properties of muscle, for these are of extraordinary interest. For the most part the voluntary muscles of the body act as sources of power, for moving the various bones, to which the muscles are attached. A tendon is chiefly composed of a bundle of white fibres intermingled with cells. It is attached to the muscle by one extremity, and narrowing into a firm, strong cord, is securely united to the bone by the other. The ligaments are tissues, the purpose of which is to bind together the structures, entering into the formation of the joints, thus rendering the union more firm.

By a sprain we understand an overstretching or rupture of some of the elements of a muscle, tendon, or ligament, dependent upon sudden or continuous strain of the tissues. As a rule, sprain of a muscle more quickly disappears under rest and proper treatment, than a like injury to a tendon or ligament. As one might naturally expect, sprains are most commonly met with in the fore feet of horses, more especially in animals used for continuous and rapid work on hard ground.

The primary essential of treatment in all cases of sprain is rest, both constitutional and local. The animal should be placed in a comfortable box, and a mild dose of physic should be given. Three or four drachms of aloes will prove of great value, in abating the inflammatory action of the injured tissues. The diet should be cooling and laxative ; mashes, oilcake gruel, grass, and carrots, taking the place of corn and hay. In those instances where a severe sprain of a fore limb has been sustained, a shoe with high calkins sometimes appears to be beneficial, in affording rest to the structures at the back of the leg. Slings are seldom necessary, excepting in cases of severe sprain of the hind limbs, or rupture of the suspensory ligament, if the animal will not lie down.

In the case of a sudden sprain, it is our custom to see that the injured part be assiduously fomented with water at about 100° F., for an hour or two, several times daily. After each fomentation, a flannel bandage soaked in a lotion made of tincture of opium one ounce, tincture of arnica one ounce, water twelve ounces, may be applied, and again readjusted after each fomentation. When the animal begins to recover, as will probably be the case in a few days, it is our custom to apply a cooling lotion of spirit and acetate of lead, or lotion of chloride of ammonium and nitrate of potassium, the formulæ for which we have given in a previous article. Internally, one ounce of bicarbonate of potassium, may be given in the water once daily, for several days. Some practitioners recommend the application of cooling lotions from the outset ; and probably this treatment, when judiciously carried out, may be as effectual as the one we have described above. A favourite method, instead of applying cooling lotions, is that of directing a jet of cold water on to the affected parts, for ten or twenty minutes at a time. In those instances in which the animal has sustained several sprains previously, and has weak legs in consequence, this method of treatment is especially useful.

In the treatment of recent sprains, exercise, we may add, should be strictly prohibited until all pain, heat, lameness, and swelling have abated. In order to promote absorption of the effusion, hand-rubbing, and pressure by means of an elastic bandage, uniformly and carefully, but not too tightly applied, will prove useful. When the animal commences work again, it is advisable to continue the hand-rubbing, should any fulness appear in consequence.

Having now concluded our general remarks, we may proceed to consider in detail the nature and treatment of the various special kinds of sprain.

SPRAINS OF THE SUSPENSORY AND CHECK LIGAMENTS.

THE suspensory ligament of the fetlock is a long and powerful brace, composed of fibrous tissue, and often containing bundles of fleshy fibres in its texture.—(Chauveau.) It is situated behind the canon bone, and between the two splint bones. At its origin from the head of the canon bone, it is quite thin ; but it soon becomes enlarged, and at the lower part it divides into two branches, which are attached to the two small bones at the back of the fetlock. The branches unite together again in front of the joint. The purpose of this ligament is to limit the degree of extension of the fetlock. Many horsemen are acquainted with the site and appearance of this ligament on the side of the leg, between the tendons at the back, and the canon bone in front. It is not certain whether it is elastic or not. Probably it is not. In a well-formed animal it is seen to stand out boldly and unmistakably, more especially in well-bred horses. Often it is obscured by the presence of additional tissue, and this is more especially the case in heavy draught horses. Although this ligament may be sprained at almost any point, the usual spot where the injury is inflicted, is at its division into the two branches above spoken of. The tendons at the back of the leg are two in number. They pass from their insertions into the muscles above, downwards behind the ligament we have been speaking of. They are termed the perforated flexor, and the perforating flexor tendons respectively. The former more superficial one divides, and the two portions are inserted into the two sides of the small pastern bone. The latter or perforating tendon passes between the two branches behind both pastern bones, over the navicular bone, and is attached to the back of the coffin bone. This is the tendon often involved in the disease of the navicular bone. The check ligament is a powerful band originating from the head of the canon bone, and becoming firmly attached to the perforating tendon, about midway between the ends of the canon bone.

It is worthy of note that, whereas sprain of the suspensory ligaments is of more common occurrence among horses used for fast work, especially when galloping on hard ground, sprain of the check ligament is more frequently met with among heavy draught horses. In the latter, however, this accident is by no means common, and, when it does occur, it generally happens in descending a hill with a heavy load behind. Those cart horses with oblique elongated pasterns are more subject to this injury. Of the tendons the perforating is the one most subject to sprain. This tendon is not uncommonly strained, owing to fast riding over heavy country, whereas both the ligaments and this tendon are often injured, in going down hill with a heavy weight behind, or in galloping. It should be remarked that, when the progression is very fast, sprain most generally affects the suspensory ligament of the fore leg. Contrary to what has been observed in cart horses, it has been noticed that in racers with elongated and oblique pastern bones, there is less risk of spraining the suspensory ligament. With upright pasterns, there is of necessity great risk of spraining the ligament, more especially when descending a hill. Racing men do not view with equanimity the risks run by such an animal, should the course have any sharp descents.

Sprain of the ligaments, one must bear in mind, is much more liable to occur when the muscles are exhausted, or are in an atonic or weak condition. Animals which often sprain the suspensory ligament, when used for racing, or for long journeys, seldom or never do so, when worked less severely, or when intervals are allowed, in which the muscles may recover themselves. Every time a muscle contracts, there is waste ; and, when the tissue is in a weak condition, it cannot so readily repair the loss. Necessarily any animal is more liable to a sprain if he treads accidentally on uneven ground, or comes upon a very hard or irregular surface, after taking a fence.

As we stated above, the suspensory band may be slightly or severely sprained, or it may be ruptured, and these injuries may affect one or both sides of the ligament. The lameness occasioned is proportional to the degree of the injury, but is always very great in severe sprains. If the ligament be quite ruptured, there is a complete break down, and the toe turns up. Heat and tumefaction follow the injury. Should rupture of the ligament involve one branch only, the resulting injury of the fetlock is not so marked.

We have seen many instances of sprain of the check ligament, which have generally been confined to cart horses ; but, of course, sprain of this ligament may occur in any horse, if he tread suddenly on a stone, or on any uneven surface. This ligament is more liable to sprain in ascending a hill, especially when drawing a heavy weight up a steep incline. When the check ligament is injured, swelling of the tissues is occasioned at the back part of the leg, between the knee and the fetlock. Heat, pain, and great lameness are additional symptoms of this accident, although, when the injury to the ligament is slight, the progression may be but little impeded or altered. The inflammatory action in these sprains is not uncommonly very marked, and may leave permanent thickening, at the point where the check ligament joins the tendon. In very slight cases of sprain of the check ligament, all that one can observe is fulness at the back of the leg, below the knee, attended with heat and tenderness. Lameness may be present, but in such cases it is rarely severe. When the tendons at the back of the fetlock are sprained, there are pain, heat, swelling, and lameness, which will vary in degree. The treatment of these sprains is that which we have already described, but we may add a few necessary details concerning sprain of the suspensory ligament. After this accident, our object is to promote union of the severed fibres of the ligament. Absolute rest is the first essential. The hollow of the heel should be well padded up with lint or tow, which must be retained there by the application of a bandage, carefully and tightly wound around the injured member. Around this, another bandage may be applied, so as to support the limb still more firmly. The opium and arnica lotion may be applied as a fomentation, and will prove serviceable in assuaging the pain, and diminishing the inflammatory action.

Sprain of the suspensory ligament is always a serious injury. With very careful management, however, and prolonged rest, the animal not unfrequently is enabled to do moderate work. This injury almost invariably leaves some mark of its presence ; and it must be considered an unsoundness, because, if the work happen to be at all severe, or there be any unusual strain, great lameness is soon developed.

CURB.

By the term curb, we understand a sprain of the ligament situated at the back of the hock joint, which makes its appearance as a swelling, as a rule some five inches from the point formed by the bones of this joint. Our readers will perhaps be aware, that the hock joint corresponds with the ankle of the human foot ; and that the bone, which forms the projection at the back of the joint, is the so-called os calcis or heel bone. Now, it is this ligament which keeps this bone in its place, and extends downwards to be fastened to the bones below, which is sprained in curb.

Of the causes of curb we have not much to say, but may mention that it usually results from leaping or galloping. Heredity, it is clear, has oftentimes something to do with the predisposition, which some animals have to sprain this ligament of the hock. This, no doubt, is to be attributed to the fact that the conformation or build of the bones and ligaments of the joint, having more especially a tendency to sprain, is inherited by the offspring. In this connection we may add that it has been said that an animal with a long heel bone is more liable to contract curb ; and this is possibly correct. In any case, the practical conclusion to be drawn regarding these facts and probabilities is, that animals which have had curbs should not be employed for breeding purposes.

As our readers are no doubt aware, a curb constitutes unsoundness, even though the progression of the animal be not altered. It is therefore very important that one should be able to detect the presence of such an injury, if it exist. The observer, in examining a horse for curb, should view the hocks at a side glance, from the off as well as the near side, carefully scanning it from above downwards. The line from the angle of the hock downwards should be straight, and should have no swelling or bulging in its course. If there be no swelling nor any alteration in the gait of the animal, the absence of curb is proved. One must bear in mind that there are sometimes noticed, enlargements of the hock, which are not unfrequently mistaken for curb. In such instances, however, on examination, it will be found that enlargement does not interfere with the progression of the animal ; and, moreover, it is sometimes present in an equal degree in both limbs. Such a condition of the hock is due to unusual size of the bone termed the cuboid, which is situated below the heel bone. Again, sometimes an enlargement is apparent when one views the hock from the outer side, but is not noticeable when one looks at it from the inner side. Such an abnormality is due to a greater development than usual of the structure, termed the external splint bone, situated below the cuboid bone.

Although, however, it thus appears that there may sometimes be some uncertainty regarding the presence or absence of curb, decision as a rule is by no means difficult. It not unfrequently happens that horses bruise their hocks at the usual seat of curb, and thus cause some superficial swelling. Advantage is sometimes taken of this fact by unscrupulous dealers, who may endeavour to persuade the purchaser that a curb is in reality a mere bruise caused by some external injury.

When the progression of the animal is not affected by the presence of the curb, even though he be employed for constant work, he is practically sound. Nevertheless, it should be remembered that lameness may result at any time, if the animal be worked hard, more especially on irregular ground, or be regularly run in the chase. This is still more likely to be the case with young animals, and we may mention that curb in a young horse is always to be regarded with suspicion, and as constituting inefficiency or practical unsoundness. In older animals of six or seven years, the tissues may be so far repaired as never to contract sprain again ; but in young ones, laxity of the ligamentous tissues and repeated sprain, may be an almost continual source of lameness.

In treating curb, the limb should be put at rest by placing a high-heeled shoe on the foot. Although it is a common custom to apply blisters immediately after a horse has "sprung" a curb, this practice must be strongly condemned. We have seen so much permanent damage done by this means, that we wish to draw the special attention of our readers to the harm it so frequently does, in the early stages. It is the best practice to apply cooling applications, until all inflammation has ceased. This may be best accomplished by directing a stream of water from a hose on to the affected hock, thrice every day. It is well to give the animal a mild dose of physic in the first instance, and to feed him upon mashes and warm water for three days. When the inflammatory action has ceased, the blistering ointment may *then* be applied. In very severe cases it is best to fire at once, and not to try blistering first.

SPRAINED BACK.

SPRAINED back is an injury not very uncommonly met with in the hunting field. It is due to sprain of the so-called psoas muscles, contracted in taking a fence, but more frequently by what is often termed "slipping up." An animal so injured can stand fairly well, as a rule, though sometimes he is unable to regain his feet. Our readers will be aware that, if the back be broken, the animal would be absolutely unable to stand, even when raised up. In these cases a dose of physic should be given in the first instance, and the animal should be fed on laxative diet, such as mashes and oil cake gruel. Locally, the sprained parts should be treated by the application of warm water fomentations, followed by the use of anodyne lotions of opium and arnica, formulæ for which we mentioned above.

SPRAINS OF THE FETLOCK, HOCK, SHOULDER, AND ELBOW JOINTS.

WE may now consider sprain of the many structures, liable to injury in the fetlock joint. We need not enter in detail into the distinctive characteristics of sprain of the separate structures of the fetlock, but may speak of them collectively. The fetlock joint when sprained is hot, swollen, and tender, and if moved causes pain to the horse. In such cases, it is well to put on a high-heeled shoe, in order to enable the animal to keep weight off the heel, and to treat the injured parts as we have already directed in our last article. In very aggravated cases, the application of ointment of biniodide of mercury is sometimes ordered. When the fetlock is not acutely inflamed, but is "full and puffy," cold applications, succeeded by the use of an elastic bandage will be very beneficial. When the ligaments of the fetlock joint are much worn, and relaxed in consequence, an unsightly bending, termed "knuckling over," is occasioned.

Sometimes a swelling of the fetlock appears at the back of the joint. It is due to an inflamed condition of the bursæ or lubricating pockets, situated between the tendon at the back and the sesamoid bones. It is hard, thus differing from windgall, which is soft. Like windgall, however, it can be pressed from one side of the fetlock to the other. This inflammation sets up serious lameness, which usually recurs on working, and is difficult to cure radically.

Of "sprung hock" we must now say a few words. By this term we understand a sprain, necessarily a very severe one, of those ligaments which bind the bones of the hock together, and of that which envelopes them in a capsule, as it were. This injury causes great lameness and pain, and the animal manifests constitutional disturbance with febrile symptoms. Extensive tumefaction appears above and below the back, and inner part of the hock joint. The general directions already mentioned will suffice for the management of this injury, which, it must be mentioned, being of a very severe nature, will necessitate six months' rest at least, and sometimes even more.

We have still to speak of sprains of the elbow and of shoulder sprain. In elbow sprain there is, as in other forms of sprain, pain, heat, and swelling of the affected parts. The limb at each step shows itself unable to support any weight, and the animal therefore drops, as it were, and is in danger of falling. Regarding treatment, the methods already indicated will suffice. When the muscles of the shoulder are sprained, an accident by no means uncommon, especially in young animals employed for drawing heavy weights, or for ploughing, they waste, and losing their tonicity or healthy condition, allow the head of the humerus to bulge out, as the animal walks. This condition is often spoken of as shoulder slip. Wasting of the muscles as has been pointed out, involves only those of the outer side of the shoulder ; whereas the wasting which sometimes ensues, as the result of chronic lameness of the foot, invades all the muscles of the leg. The progression of the animal in shoulder sprain is diagnostic of the seat of injury. The injured limb is not brought directly forwards, but it is moved in a kind of rotatory manner. The toe thus describes part of a circle, and is drawn or dragged along the ground. Inflammation of the shoulder joint and other injuries of this part give rise to the same characteristic dragging of the foot, and rotatory motion of the limb. The animal shows no sign of pain in the foot, nor does he manifest any, in putting it to the ground. When, however, he brings it forward to the front, he exhibits signs of hesitation, and suffers pain. The treatment of shoulder sprain consists of warm fomentations, followed up shortly by the application of smart blisterings. Rest is essential, and the application of a high heeled shoe advisable.

CHAPTER V.

FRACTURES AND DISLOCATIONS.

INASMUCH as there is often great difficulty in treating fractures of the bones of the horse, owing to the fact that absolute rest is not easily maintained, we shall not enter into a very elaborate discussion of this subject. A fracture is spoken of as simple, when the bone is broken at one spot only, and when there is no external wound, extending down to the seat of injury. When such an external wound exists, the fracture is termed compound, and when the bone is broken in several parts, it is termed comminuted. When a bone of a limb is broken, lameness necessarily results. The animal often manifests great pain, and displacement of the parts is often noticeable. If the broken ends are rubbed together, a grating sound may often be heard. In some instances, there is no distortion of the normal relationship of the tissues.

We purpose to say firstly a few words regarding the general treatment of fractures, and then to describe some of the more common forms in detail. Compound and comminuted fractures are always difficult to treat, far more so than simple ones.

In the treatment of a fracture, absolute rest is the first essential; and, in order to secure this, slinging will in many instances be indispensable. When the fracture is compound, it will be necessary to treat the wound as well as the fracture by the application of some antiseptic lotion, as for instance of carbolic acid, water, and glycerine ; and, moreover, all detached fragments of bone should be carefully removed. The question of the advisability of treating a fractured bone in a horse is often a pecuniary one. The requisite food, rest, and attendance, skilled and manual, are serious items of expense ; and moreover the chance of failure, owing to the difficulty of maintaining the broken ends in close apposition, is a point for due consideration The broken ends of a bone are brought together, and secured thus, by means of splints made of gutta percha or leather. Sometimes, bandages moistened with hot water, and then covered with plaster of Paris, are employed. The plaster sets firmly, and the fractured bone is thus rendered firm and secure.

Sometimes what is spoken of as a charge proves very valuable in maintaining the broken bone at rest. By a charge we understand a cotton bandage about four or five inches in breadth, on which is placed some material which sets hard. Such a preparation may be made of equal parts of ordinary pitch and Burgundy pitch. This method of treatment is to be preferred to that of securing apposition by means of plaster of Paris. In the case of a limb, when it is purposed to employ splints, one is placed on each side of the injured member ; and then a bandage covered with plaster of Paris or starch is wound not too firmly round the whole. It is well, before adjusting the splints, to place tow or lint around the injured limb, so as to fill up the gaps and irregularities of the surface. When there is an external wound, this must be left exposed to the air, and thus an aperture corresponding with the open injury must be left in the splint. In most instances, it will be necessary to allow the splints or charge to remain in place for six to eight weeks. At the end of this time, they may be removed, and bandages should then be firmly applied. The animal cannot be exercised, until after the lapse of at least sixteen to eighteen weeks, after sustaining the injury. If the animal manifests great pain, an ounce or two of tincture of opium may be administered. During the treatment of the injury, the bowels should be regulated by the administration of an occasional dose of physic ; and the animal should be fed on a nutritious laxative diet, consisting of oatmeal gruel, grass, and carrots.

The fractures we most commonly meet with are those of the pastern bone, skull, thigh bone, tibia, and back. Fracture of the pastern bones generally occurs as the result of hard and fast riding and galloping, over irregular ground. Sometimes a pastern bone is broken in one part, and in other cases in several. The long pastern bone is more often fractured than the short. This injury, contrary to what might be anticipated, is not in every instance attended by marked signs. Lameness, however, in most cases is very pronounced, and the poor animal is not able to bring his foot down to the ground. Distortion of the parts may or may not be manifest, but pain and swelling are generally present. In those instances where the bone is broken in several places, treatment is generally not successful ; but when only broken in one place, and when little or no displacement occurs, recovery is to be expected. At the same time it may be mentioned that an animal so injured is, after recovery, rarely fit for fast work again. The animal should be placed in slings in the first place, and the shoe should be removed. The best method of treating the injury is to fill up the hollow behind the pastern with tow charged with pitch, and then to wind a narrow bandage nine or ten feet long similarly charged around the limb. When the bone has united, as it probably will have done, in the course of about five or six weeks, the charge may be removed ; and, if there be much swelling, owing to the new bone thrown out, the part should be smartly blistered.

By broken back, we understand fracture of one of the vertebræ or bones of the back, a most serious injury, generally caused by a violent fall. Sometimes, as we mentioned in treating of sprains, broken back is difficult to distinguish from sprain of the muscles of the back. The former is necessarily of far greater danger, and, though often a remediable accident, when the fracture involves one of the vertebral bones, from which the ribs extend to encircle the chest cavity, it is nearly always fatal, when the column of bones is broken in the region of the loins. In the latter case, the paralysis, caused by the broken bone pressing upon the spinal marrow, is often absolute ; while in the former case the paralysis is mostly not so severe. In broken back, if the animal has not regained his feet, and it is deemed advisable to give him a chance of recovery, he must be treated in the recumbent position. If he regains his feet, he should be placed in slings. When the bones are much displaced, and great irregularity is felt in passing the hand down the back, and when paralysis is very marked, recovery is mostly impossible, and it is the kindest course to put the animal out of suffering.

Of the dislocations found in the horse, the two most common are dislocation of the patella and of the shoulder. We purpose here only to speak of the former. The patella—the little bone in front of the stifle joint of the horse—corresponds with the knee-cap of the human being. Foals, more especially weakly and debilitated animals with lax tissues and ligaments, are greatly liable to this injury. Dislocation of the patella may be partial or complete. It is an easily recognised injury. In cases of complete dislocation, the limb is held stiffly, and is directed backwards. The front part of the hoof, moreover, is brought into contact with the ground ; and the animal walks with great difficulty, the limb being carried stiffly and straightly. There will also be lameness when the injury is only partial, but this will not be so marked. In partial dislocation, the bone will sometimes pass back into its proper position, when the animal is suddenly moved forward.

The best method of reducing dislocation of the patella is to tie a firm rope round the pastern bone of the injured leg. The rope should then be pulled forward by one man, while another individual should press upon the displaced bone, and endeavour to push it into its proper place. While this is being done, it is best to back the animal. Sometimes this method is not successful. In this case, chloroform is often administered, with a view of causing laxity of those muscles, which are attached to the patella. After the reduction of the patella, it is necessary to take steps to make it retain its proper position. With this object, the rope attached round the pastern is passed forward round the neck as a collar, or it is attached to a collar. A blister may now be applied to the limb, and may be repeated in about a week's time. It is well to be careful at first not to allow too much exercise, but to increase it gradually, as the animal is able to bear it.

CHAPTER VI.

SPLINT, BONE SPAVIN, SORE SHINS.

SPLINT.

By a splint, we understand a bony deposit, formed as the result of inflammatory action, generally on the upper and inner third of the cannon and splint bones of the fore leg. In some instances, the osseous deposit may be formed on the outer side of the leg, and in other cases it may involve the outer as well as the inner side. The bony growth, moreover, may be found midway between the inner and outer aspect, and instances in which splints have been formed on the hind limbs are not very uncommon. In the latter situation, they generally occupy the outer aspect of the canon bone, and very rarely cause lameness.

There are five classes into which splints may conveniently be divided :—

Firstly, simple splints. By a simple splint, is meant a deposit of bone, which does not interfere with the tendons and suspensory ligament, and is situated at a distance from the knee. Simple splints, when not impeding the progression of the animal, are not to be regarded as constituting unsoundness. All other kinds of splints, and simple splints when causing lameness, as they sometimes do, in the early stages more especially, constitute unsoundness. It may be mentioned that this variety of splint more often provokes lameness, when seated on the outer side of the leg.

Secondly, double or pegged splints. Splints are termed pegged, when there are two deposits, one on the inner and one on the outer side, connected through the leg by a bony communication.

Thirdly, splints situated near the knee.

Fourthly, two deposits, one above the other, on the same side of the leg. Sometimes there is a bony connection between them.

Lastly, bony deposits involving not only the splint bones, but a bone of the knee joint also.

Concussion is the chief cause of splints. Heredity also is a potent factor in predisposing to the formation of these inflammatory deposits. Splints commonly owe their origin to the fact that the animal has been trotted on hard ground, more especially when this has been kept up continuously. Young animals, particularly when first put to work and too heavily weighted, frequently develop splints. Our readers will be aware that splints often appear, while the young animals are as yet capering in the grassy fields, unbroken and untouched by the hand of man. Not only, however, are horses of under five years more subject to splints, but at this period of life they are also more commonly rendered lame by these bony deposits. When one learns that concussion is the chief cause of splints, one will readily see how it is that the more purely bred animals, since their work is faster, more frequently develop splints, than animals of coarser breed. One will also understand that a splint in a draught horse, not required for fast work, is of less serious moment, than in more rapidly moving animals. Although we attribute the greater immunity from splints, which cart-horses enjoy, more especially to the smaller amount of concussion, which their legs undergo in progression, we must bear in mind that their limbs, being of much heavier build, are not so liable to be injured by continual shock, as are those of finer bred horses. We do not often meet with splints in old horses, but they are sometimes seen, even in horses of advanced age.

The bones of man and most animals are covered over by a fibrous envelope; and it is inflammation of this covering or periosteum, as it is termed, as well as of the bone itself, which results in the formation of those osseous deposits, which we know under the term splints. One can easily imagine that a young bone, not fully grown, is more liable to become inflamed by work, more especially when such work causes much concussion.

It is fortunate that a great number of splints are those which we spoke of as simple; and, as these rarely cause lameness, excepting sometimes during their period of growth, that is in the early stages of inflammatory action, the progression of the animal is not so often interferred with, by this disease of the bones, as one might expect. A splint, situated on the outer side of the leg, in most instances causes greater lameness, than one seated on the inner side.

It should always be borne in mind, that lameness from a splint does not depend upon the size of the deposit. Sometimes, indeed, a very small deposit causes marked lameness, while at other times a very large bony growth may not even alter the gait in any way. One cannot gauge the amount of alteration in the bone by the external size and conformation of the splint, because a very small superficial growth may co-exist with inflammatory deposit, existing more deeply. It is not strange that horses often manifest marked lameness, as the result of inflammatory action of the bone, while as yet there is no external sign of a splint. In such cases, the amateur is often at a loss to account for the lameness. There are, however, certain facts which help us in diagnosing correctly, whether the lameness proceed from inflammatory action of the bone, or not. Lameness in a young animal, in the first place, is more likely to be due to this cause; whereas in older animals it is more likely to proceed from navicular disease.

Again, it is well known that a horse, whose progression is altered by the existence of a splint, walks sound or nearly so; whereas in trotting, the lameness is very marked. Manipulation of the leg also often reveals heat and tenderness at the usual seat of splint, and a small hard rounded growth may sometimes be discovered. Pressure upon the site in such instances, may show such tenderness, as to cause flinching. We have met with instances, where no deposit was thrown out for some length of time after lameness became manifest. Sometimes in splint lameness, there is diminution of the proper amount of bending at the knee, as the horse moves. Our readers will remember that we said that horses, suffering from navicular disease, as a rule improved in their action during exercise. In animals with splints, however, the lameness as a rule becomes more marked during progression. This, we may remind our readers, is also the case with horses having corns. When one wishes to examine a leg, with a view to detecting the presence of a splint, one should grasp the limb in the usual manner, with the fingers upon one side, and the thumb upon the other, and then should trace the splint bones from above downwards. Should there be any growth, it will readily be felt.

In those instances in which the splint does not cause lameness, it is customary not to interfere with the disease. The animal should be put upon a diminished diet scale, and his food should be of a laxative nature. It is well to give an aperient, and afterwards enjoin that no exercise should be given. These injunctions should be ordered to be carried out, until the inflammation has ceased. In those cases where the lameness is not very marked, it is best to rest the animal for a time, and blister the inflamed bone with ointment of biniodide of mercury. A dose of aloes should also be given, in order to lessen the inflammatory action. Should the blister not prove curative, it will be necessary to fire the part with the prick-iron. When the lameness produced by a splint is very severe, and the animal places but little weight on the limb, Mr. Sewell's operation of periosteotomy is sometimes performed.

BONE SPAVINS.

WE may now turn our attention to the consideration of spavin. Few diseases of the horse are so commonly before our notice as spavin, and few cause so much litigation, and give rise to the expression of such diverse professional opinions. Regarding the origin of the word spavin there is also considerable doubt. The Latin word was employed by Jordanus Rufus, in the thirteenth century; but we cannot say whether he originated the term, or not. Some writers believe it is derived from the Italian *sparavano*. Others again derive it from the Greek word *spasmos*, a spasm or cramp. Winter derives the term from the French *ésparvin*, while others again believe it to have its origin from the Latin *sparsus*, on account of the straddling gait, which often results in this disease of the hock.

A spavin may be defined as a deposition of bone on the inner and lower part of the hock, resulting from chronic inflammatory action of certain bones composing this joint, and generally resulting in their cementing or anchylosing together. Our readers will understand that the bones affected by spavin are not those forming the true hock joint; but are the canon bone and the little bones situated just above it. Sometimes, we may add that the bones of the true hock joint do become implicated, as the disease spreads ; but it is not by any means a common occurrence. One will readily understand that the higher the bony growth is deposited, the more grave are the consequences. On the other hand, when seated lower down between the canon bone and the little bone immediately above it, the cementing together of the joint is not of great moment. The progression of the animal in this case is not much affected thereby, as there is but little motion in this joint. The causes of spavin are of two kinds, actual or external, and predisposing or internal.

There are a number of predisposing causes, which we may briefly consider. Firstly, the bones of young and overgrown horses being soft and immature, are more liable to become diseased. One can readily understand that, at an early age, the bones and their coverings, as well as the joints and ligaments, are most liable to become diseased, especially when the animal is heavily weighted, before his bony tissues are really consolidated. Secondly, irregular conformation of the hock joint is also to be regarded as a predisposing factor in the causation of spavin. When the angle of the hock is less than 135°, the animal is termed " sickle-hocked." Such a conformation of hock we mentioned was unsuited for rapid progression, and is especially liable to curb. Likewise it is believed to be more liable to become the seat of spavin, though the predisposition to this disease is not so marked, as in the case of curb. Wide hocks, sometimes caused by bad shoeing, or disease of the fetlock joint ; straight hocks, in which the angle of the joint is more than 160°; and cow hocks are also examples of irregular construction of these parts. We do not know that these latter irregularities increase the liability to spavin. Animals with "laced-in hocks" or "tied in below the hock" are also especially subject to spavin ; and this is also the case in animals with short or round hocks. Thirdly, excitable animals of irritable temperament, it is believed, are also more subject to contract spavin. Fourthly, animals with long backs and narrow hind quarters, are more prone to the osseous disease of the hock in question. Lastly, hereditary predisposition is very marked in this inflammatory disease.

The actual causes of spavin are strain or concussion of the structures of the joint, due to galloping, or very hard work, or wearing shoes with too high calkins, and imperfect food supply. Necessarily, if the food is insufficient and of bad quality, the animal will be more subject to spavin, and, indeed, also to sprains of the various muscles and tendons of the body. It is in cart-horses that high calkins not uncommonly are answerable for the production of spavins ; and this is especially the case, when the animal is employed for drawing heavy weights down hill. It is necessary that animals employed for hunting should have well formed hocks, because the amount of concussion on the hocks in leaping renders them more liable to suffer from spavin, and one rarely sees an animal which has been hunted for a couple of seasons without observing that he has thrown out spavins. It is not strange that the more forward the spavin appears, the greater is the resulting impediment to the progression.

O

Coarse hocks are not necessarily to be regarded unfavourably, as they commonly become finer, as the young animal matures in age. If, however there be any dissimilarity in the size of the two hocks, or if there be any lameness, our suspicions are aroused. If the hocks are coarse in an old horse, and there is no alteration in the animal's gait, lameness very rarely indeed results. It is fortunate that in horses six years old and under, spavin is generally amenable to treatment. In old animals therapeutic measures are of little avail.

Spavin lameness is sometimes difficult to distinguish from alteration of the progression owing to other causes. Hip lameness, we should remember, causes stiffness of the whole limb ; and there are generally tenderness, heat, and swelling at the hip, in such cases. The toe also is dragged along the ground in hip lameness. Regarding the judgment we may pass upon the spavin, we may point out that animals with well-shaped hocks, as a rule, more speedily recover, and are less likely to have a renewal of the inflammatory process, than others with badly-shaped hocks. Spavin also is more damaging to an animal required for fast and heavy work, than for one employed for lighter work at a slower pace. A spavin situated internally is, as we have pointed out already, less likely to interfere with the progression, than one situated on the front of the joint ; and low spavins, we also mentioned, were less likely to cause lameness than high ones. Moreover, when the true hock joint is implicated, there is but little chance of recovery. Bony deposits, situated at the back part of the inner surface of the hock, do not often cause much lameness. In animals well shapen, a spavin is not likely to be so damaging as in long-backed horses, badly ribbed up, having poor appetites. When the deposit is associated with string-halt, or causes any other disease by its presence, the prognosis is not so favourable.

In those instances where there is no external sign of any bony deposit, even although there may be very marked lameness, the disease is spoken of as occult spavin. In occult spavin, no bony deposit is thrown out, but the inflammatory action results in ulceration of the contiguous surfaces of the bones, and is not followed by any reparative process. Our readers will understand that spavins, like splints and other such bony growths, are in reality "nature's means of fortification against more serious failures." In occult spavin, this reparative process of bone formation, in order to make good the loss by ulceration, does not occur ; and therefore there is no external sign of the disease. Occult spavin is more grave and intractable than ordinary spavin. It is of much more frequent occurrence in old animals than in younger ones.

We may now say a few words regarding the means of detecting spavin, and then proceed with the treatment of this common disease. In spavin lameness, there are some points specially noteworthy. There is a lack of bending at the hock joint. The lameness, as a rule, is less marked, after the animal has been exercised for a time. The step will be noticed to be on the toe, at which part the shoe is consequently more worn, than elsewhere.

A horse with a spavin, as Percivall said, is especially lame on stepping out of the stable, on the day following after a heavy day's work. Dragging of the toe is sometimes noticed, on riding a horse down a steep hill; and, in this manner, spavin has been not unfrequently detected. Percivall was the first authority to whom we are indebted for accurate descriptions of spavin. Regarding the actual position generally taken by spavin, we cannot do better than quote his words. "It is precisely the interval between the prominence where the hock ceases, and the canon-bone begins, that is the site of spavin. A small round tumour interrupts the natural declivity from the hock to the canon bone, and in a moment catches the eye of the experienced observer. In cases where the tumour being small, or flat, or diffuse, is indistinct to the eye, the observer will not make up his mind concerning it, until he has narrowly compared the suspected, with the sound hock."

In some severe cases, the lameness of spavin is characterised by a kind of spasmodic jerking up of the limb, at the instant the heel comes into contact with the ground. Sometimes there may be no lameness; but, when active change is going on in the bones, this is rarely or never absent. In all cases the examiner should feel both hocks, when he wishes to compare them with the view of noticing any difference of conformation. It is customary to examine the near hock with the right hand, and the off one with the left. Often a spavin can be felt, when it is too small to be observed with the eye. Animals with spavin should not be used for breeding purposes, unless, indeed, the disease be due to some external cause, such as a sprain.

In those cases of lameness from spavin, which are seen in the very early stages, a purgative should be given in the first instance, and the animal should be rested, and shod with a high-heeled shoe. Locally, a blister of ointment of biniodide of mercury may then be applied. If ineffectual, setoning or firing will then be necessary. Unless the spavin gives rise to lameness, treatment is seldom carried out. Firing is commonly adopted in those chronic cases which have resisted the milder remedies, such as blistering, or the douching with cold water, practised by some. In applying the actual cautery, the horizontal lines should not be made too closely together, but they should be pretty deeply burned in, in order to act more effectually. After firing, a rest of six or seven weeks or more is necessary, before the animal is again fit for work. Prick firing we frequently advise, as it often proves more valuable than stripe firing. The prick iron we use is different from the one commonly in use; the prick at the extremity being only about one-sixth of an inch in length. This is heated red-hot, and is then thrust in in several places (see page 208). We do not recommend setoning in cases of spavin.

If it be decided to pass a seton, it is best to have the animal cast, and one must take care not to injure the large vein, which runs over the inner and front portions of the hock. The seton is passed vertically exactly over the growth, or we may insert two smaller ones on each side of it. After setoning, the animal should be rested, and fed for a fortnight on a cooling laxative diet, and a shoe with high heels should be applied. After the wound is healed, the part may be smartly blistered. Some horsemen are very fond of corrosive applications, such as those composed of corrosive sublimate, dissolved in spirit; but these, although often useful and efficacious, sometimes destroy the skin, and thus cause a blemish.

SORE SHINS.

By the term "sore shins," we signify a disease of the canon-bone and its covering, usually affecting its lower and front part, and only differing from a splint, in that it affects another portion of the bone. In some instances, the affection involves the whole length of the canon-bone. Like splint formations, this disease also is mainly due to concussion, and is likewise especially common in young animals, worked before the bones are matured. Sometimes all four canon-bones are diseased; but, as a rule, the affection is present only in the fore legs, as these necessarily are more liable to suffer from shock, than the hind ones. It has been noticed that the leg, with which the animal leads in the gallop, is more often affected, than the other ones. Sore shins are rarely met with excepting in young race-horses, among which it is a common disease.

The lameness occasioned by sore shins is insidious. The animal is at first observed to step somewhat short. Unless the disease be arrested, as soon as the first symptoms appear, by rest and treatment, the lameness becomes very marked indeed, and constitutional symptoms are manifested. When the inflammation is severe, the pulse is increased, and swelling at the lower third of the canon bone appears. When handled, this will be noticed to be soft, elastic, and very tender to the touch. In very severe cases, unless active measures be taken, necrosis or death of the bone may ensue, and the disease may even prove fatal. In less extensive disease, the swelling is circumscribed, and afterwards becomes quite hard.

In cases of sore shins, a purgative should be given, and the diet should be laxative. If the disease is not very severe, cold applications should be assiduously applied, and these should afterwards give place to a smart blister. In very severe cases, it is necessary to cut through the inflamed covering of the bone. Fomentations of warm water, with applications of antiseptic lotions (carbolic acid one part, water twenty parts), are then necessary.* As the inflammation subsides, blisters are called for. In cutting down upon the canon-bone, the veterinary surgeon must be careful not to injure the tendons.

When the disease is cured, it is very essential that the animal be rested for a time, and then put into work gradually. The exercise at first should be on soft ground, and the animal should not be allowed to gallop much until he is well able to bear it.

---

*Mr. Charles Gresswell, of Nottingham, who has had large experience in the treatment of sore shins, especially recommends the ointment of boracic acid.

CHAPTER VII.

BURSAL ENLARGEMENTS.

Bog Spavin. Windgalls. Thoroughpin. Capped Hock and Elbow.

THOROUGHPIN.

THOROUGHPIN is a bursal enlargement, which appears as a swelling on the lower and lateral aspect of the thigh, at the upper and back part of the hock. It is due to a swollen condition of the sheath, which envelopes the flexor perforans tendon ; and this may be owing to disease of the tendon itself, or to disease of the sheath. If a thoroughpin be pressed upon, it may be made to move from one side to the other. In size these bursal enlargements differ very much, varying from that of a pigeon's egg to a child's head. It has been noticed that they are more commonly found in animals with short vertical hocks ; and that heredity has much to do with the predisposition, which some horses have to this form of disease. Thoroughpin constitutes unsoundness, although fortunately it is in almost all cases a curable affection. Generally it is consequent on sprain of the tendon, sustained by moving a heavy weight ; though occasionally it is due to an over-secretion of fluid in the sac. In the latter case, when it is termed dropsical, it is as a rule more amenable to treatment, than when following a severe sprain.

It is our custom, in treating thoroughpin, to order the attendant to direct a flow of water from a hose, for half-an-hour three times daily, on to the swollen hock, and to apply a well-fitting spring truss, which is easily procured at a small expense. This method of treatment is nearly always successful ; but the length of time it must be persevered with, varies greatly. When it is not efficacious in causing disappearance of the thoroughpin, it is our practice to blister the swelling with equal parts of ointment of biniodide of mercury and of cantharides. From the first in all cases it is advisable to apply a high-heeled shoe. Should this not prove successful, firing with the stripe-iron will always prove efficacious in reducing the swelling. Some writers recommend the application of the ointment of iodine, or of the liniment of this drug, in preference to the cold water treatment. The method of treating thoroughpin, by evacuating the sac by means of a puncturing needle, applied carefully under the skin, we do not recommend, as it is, to say the least, dangerous, and in many hands has proved fatal. Just recently we have had under treatment a thoroughpin nearly as large as a child's head. This disappeared after firing, and has since shown no signs of recurrence.

WINDGALLS.

By the term windgalls, horsemen understand those small puffy swellings, commonly met with in different positions on the fetlock joint. These little swellings, which are due to a distention of the synovial sacs of the fetlocks, are not of very serious moment. When the windgall is situated between the tendon at the back of the fetlock joint and the sesamoid bones, it is spoken of as thoroughpin of the fetlock. This name is given to it, because, being prevented by the tendon from projecting backwards, it makes its appearance on both sides as a divided swelling at the back of the joint. The treatment of windgalls is not of much importance. The application of a bandage, tightly applied, will often prove useful by maintaining firm pressure on the distended sacs. Rest also tends to reduce them. Mild blisters, such as liniment of iodine, are in most instances only temporarily efficacious. The cold douche, followed by the application of bandages moistened with some cooling lotion, often proves very useful by astringing the relaxed tissues of the part. The treatment, however, which we recommend is the application of a smart blister.

BOG SPAVIN.

By the term bog spavin, we understand an elastic, boggy swelling, situated at the inner side and front of the hock joint. It is a distended condition of the synovial or lining membrane of the true hock joint; and it occupies therefore a higher position than that generally taken by an ordinary bone spavin, with which it has no relationship except in name. As it increases in size, it extends up the leg for several inches. Bog spavin is of two chief varieties, which must be carefully distinguished from each other ; as, while one kind constitutes unsoundness, the other does not, as a rule, impair the usefulness of the animal. The former variety is caused by inflammation of the hock joint. When this is acute, there is marked lameness, and the animal cannot put his leg to the ground. The pain, tenderness, and swelling, which is hard, are great ; and constitutional symptoms manifest themselves. When the inflammation is of a chronic variety, it may be due to a rheumatic affection of the hock joint bones or other causes. When the affection of the bones of the hock is of a rheumatic nature, extensive changes occur in the cartilages of the joint, and the disease is incurable. The other variety of bog-spavin is, just as in the less severe variety of thoroughpin, due to a dropsical condition of the joint, and is not generally provocative of pain. In this case, the swelling is not tender or hot, and when felt is found to be boggy and elastic. The inflammatory variety of bog-spavin is generally due to sprain of some of the structures of the hock joint ; whereas the dropsical variety is generally due to overworking, while the bones of the animal are not full grown. Heredity, as in the case of thoroughpin, we believe, is sometimes answerable for the predisposition which some animals have to bog-spavin. Cold, wet, and chill are potent factors in the causation of the rheumatic variety.

In cases where inflammation is present, it will be necessary to give an aperient, say, four to six drachms of aloes, and to feed the animal on laxative diet. Rest is essential. Locally, cooling lotions, such as one composed of spirit one part, solution of acetate of lead one part, water eight parts, may be applied at frequent intervals. Afterwards, and in dropsical cases, from the first the joint may be blistered with ointment of biniodide of mercury, or with a tincture made by dissolving thirty-five grains of bichloride of mercury in two ounces of methylated spirit, and adding forty grains of biniodide of mercury. If this does not prove efficacious, firing is necessary, and will almost certainly prove curative, as, fortunately, all varieties of bog-spavin, excepting the rheumatic, prove almost universally amenable to judicious treatment. The fluid effused cannot be drawn off by puncturing, as it is contained in the true hock joint, which must on no account be opened. We have not seen many cases of so-called blood-spavin. By this term is signified a varicose condition of the large vein running over the inside of the hock. Such a condition is very rarely met with.

CAPPED HOCK, KNEE, AND ELBOW.

THERE are two varieties of capped hock. The commoner kind of this affection is a serous sac, situated at the point of the hock, between the skin and the tendon situated there. It is generally due to an injury, the result of a sharp blow sustained in kicking, or in other ways. If there be heat in the part indicating inflammation, cooling lotions should be applied locally ; and, when the acute stage is over, the ointment of biniodide of mercury will prove valuable, if repeated at intervals of several days. There is a bursa situated at the point of the hock ; and, when this becomes inflamed, as the result of a sprain or other injury, the second variety of capped hock is produced. This bursal enlargement, which is rarely met with, can easily be distinguished from the other variety of capped hock, by the fact that it makes its appearance on both sides of the joint of the hock, as an elastic fluctuating tumour or swelling. Although the previous variety cannot be regarded as an unsoundness, this form does render the animal unsound. The only treatment of value for this affection, is either blistering with ointment of red iodide of mercury, or firing. General rest is necessary, and local rest of the limb should be secured by the application of a high-heeled shoe.

Capped knee is due to a swollen condition of the bursa in front of the knee joint. This bursa is that over which the tendon in front of the knee plays ; and it is not unfrequently injured by blows, or by the entrance of thorns into it, when it becomes swollen, and distended with effused fluid. As soon as the injury is sustained, pain and lameness usually manifest themselves. In these early stages, rest, an aperient of aloes, and the application of warm water fomentations are requisite. As the inflammation passes off, it is well to blister the swollen part with ointment of equal parts of biniodide of mercury and of cantharides. If this does not cause the absorption of the effused fluid, is is best to puncture the distended sac, at its lowest point at the innermost part, by a horizontal incision. The fluid may then be squeezed out. The most scientific and at the same time the best method, is to draw off the fluid by the aspirator. After the application of a blister, pressure by means of a bandage should be maintained. This is also necessary after the evacuation of the sac by puncturing. In the latter case the bandage should be kept firmly applied, so as to maintain the walls of the sac in close contact, and to cause their union together.

Capped elbow, like the rarer variety of capped hock, is due to the appearance of a serous sac at the back of the elbow joint. It is generally caused by a bruise, the result of lying down with the heel in close contact with the elbow. Warm water fomentation, assiduously carried out, is the best treatment. After each fomentation, the elbow may be rubbed with the ordinary white linament. The fomentations should be carried on for an hour at a time, four or five times daily, for a few days. If these measures are not successful in causing the disappearance of the tumour, it may be smartly blistered, or a seton may be passed through it. In some instances, when the tumour becomes indurated, it may be removed by an incision in the vertical direction through the skin.

CHAPTER VIII.

Poll-Evil. Inflamed and Fistulous Withers. Open Joints. Rupture. Choking. Osteo-porosis. Melanosis.

POLL-EVIL.

By the term poll-evil, we understand the growth of one or more abscesses at the upper and back part of the skull, just behind the ears. This unfortunately common malady usually results from an injury, whether it be caused by a sudden blow, or by the use of a tight bearing rein. There is, as a rule, no difficulty in recognising poll-evil. In the early stages, one finds a soft diffuse swelling in the position mentioned, and the neck is held stiffly. The swelling is tender to the touch, more especially at first, and it gradually becomes more defined, and commonly ends in the formation of an abscess. When we see a case of poll-evil, before matter is formed, it is our custom to order the application of cooling lotions, such as those of acetate of lead and spirit, and to order the administration of a dose of aloes internally, with the view of possibly preventing the swelling terminating in the formation of an abscess. If, however, matter be already formed, it is necessary to freely open the abscess at its base, as early as possible, and to foment the surrounding parts assiduously with warm water. The wound must be kept strictly clean ; and in order to promote its healing, the application of ointment of boric acid, or of that of eucalyptus iodoform and carbolic acid, the formulæ for which we have already mentioned, will prove very useful.

It not uncommonly happens that the inflammatory process spreads, and the disease then assumes a more chronic and obstinate character. The abscess having been left to burst spontaneously, discharges, in such a case, an unhealthy, fetid matter, and shows no tendency to heal. The "pus' burrows in various directions among the ligamentous tissues of the neck. In treating an animal thus afflicted, professional skill is necessary. It is our practice to freely open the burrowing channels with the knife, and then to dress the wound once with some strong caustic solution, such as that of chloride of zinc. After thus destroying the walls of these sinuous passages made by the burrowing matter, we dress the wound with carbolic acid lotion, or ointment of iodoform and eucalyptus. Although such cases as these frequently prove troublesome, they are, as a rule, amenable to careful and judicious treatment.

INFLAMED AND FISTULOUS WITHERS.

WE must now speak of inflamed and fistulous withers. Excepting in position, this condition is closely allied to poll-evil. Its nature and causation are similar to those just considered. Bruises, inflicted by ill-fitting saddles, are generally answerable for the production of these conditions of the withers. It will readily be understood that an animal with highly-elevated withers, will be especially liable to injury from this cause. The principles of treatment of these conditions are similar to those already described in speaking of poll-evil. In slight cases of bruised and inflamed withers, the application of cooling lotions, and the administration of a dose of aloes, will generally suffice. The diet should be of a laxative nature. When matter is formed, it is necessary to open the abscess freely, at its lowest part, as early as possible, so as to prevent it burrowing among the tissues. Warm water fomentations, and the application of poultices of bran or other material, are then necessary.

There is, as a rule, no difficulty in perceiving when the inflammation has resulted in the formation of an abscess. The soft fluctuating feeling of the imprisoned matter, and the falling of the hair from the most prominent part of the swelling, indicate the formation of matter. When, as sometimes happens in poll-evil, the case assumes a chronic form, as the result of the burrowing of matter, or from portions of bone decaying, and thus causing irritation, the diseased channels must be freely opened and treated as in poll-evil. The application of a smart blister around the diseased part is sometimes valuable in such cases in promoting healing.

OPEN JOINTS.

ONE of the gravest forms of injury to which the horse is liable, is the opening of an important joint. The joints most commonly thus seriously injured, by a kick or a fall, are the hock and knee. These injuries are followed by very marked constitutional disturbance, manifested by high fever, and there is, in most cases, great emaciation, resulting from the continual escape of synovial fluid from the joint. Such a discharge of fluid is usually an early symptom, and indicates the serious nature of the injury sustained. Sometimes, we should point out, the oil may ooze from the wound of a joint, even when the latter is not opened. This occurrence is then to be attributed to the fact that the synovial covering of the bones entering into the formation of the joint is in an inflamed condition. The serous discharge from an open joint, at first thin and pellucid, soon becomes thicker, yellowish, and purulent. The joint swells more as the inflammatory action becomes more established, and will be observed to be very painful and tender. The febrile symptoms, in unfavourable cases, show no signs of abatement, the pain continues to be very severe, the appetite is wholly or partially lost, and the poor animal sinks, as the result of the constitutional disturbance and the exhausting discharge. In those instances where the injury does terminate favourably, the bones forming the joint often become united together, and the animal has a stiff joint for life.

We do not propose to enter deeply into the treatment of open joints. Considerable skill is requisite to secure a good result in such cases. The animal should be placed in slings, as otherwise he will soon fall, and be unable to rise again. A dose of aloes should be given in the first instance, and the diet should be of a nutritious, laxative nature. As the disease progresses, it is most important that the diet should be as tempting and as nutritious as possible, as the continuous discharge proves very debilitating. After the administration of the aloes, if the febrile symptoms are very marked, a drench, containing three ounces of liquor ammonii acetatis, and five drops of Fleming's tincture of aconite, may be given in a little water, every four hours, until the acute symptoms subside. There are several plans of treating the injury locally. Probably one of the best of these is the antiseptic method. The wound is carefully bathed with lotion of carbolic acid (carbolic acid one part, glycerine four parts, water twenty parts), and is then covered over with a quantity of gauze or lint of eucalyptus and iodoform. This acts as an efficient antiseptic. Around the antiseptic dressing, a bandage is then not too firmly wound. The wound will require to be dressed every other day, or more frequently, if there be any accumulation of matter. Some authorities recommend the application of a blister around the opening, when the swelling is very marked. In some instances, this plan seems to be attended with beneficial results.

RUPTURE.

OF ruptures or herniæ, as these injuries are termed, there are two kinds to which the horse is liable. At the navel, rupture through the walls of the belly is not uncommon, and is readily detected. In this kind of rupture, which is generally present at birth, or occurs shortly afterwards, the bowel may or may not escape into the tumour. In many cases the rupture disappears spontaneously, more especially in young foals. In order to cure this variety of hernia the animal should be cast, the bowels being previously opened by the administration of a pint of linseed oil. For several hours before operating, all food should be withheld. After casting the animal, the swelling is pushed up into the belly, and the loose skin is then drawn up tightly on the fingers, and maintained thus by two skewers passed through it, one at either end. The skewers are then fastened together, and the skin drawn up is held firmly, by strong twine passed round them. The skin then gradually sloughs away, and needs, as a rule, no further attention. The ligatures must not be too tightly applied. When the rupture occurs in the walls of the belly, but not through the navel, it should be passed back and retained there, by the careful application of bandages.

The second variety of rupture, of which we need not speak at length, is termed inguinal hernia. In chronic cases, as a rule, there is not much harm from this form of rupture ; but, when the rupture follows suddenly any sudden exertion, professional aid must be at once secured, as otherwise death will be almost certain.

CHOKING.

CHOKING is due to the impaction of a portion of food. Cut hay or chaff, swallowed rapidly, is especially liable to cause this condition. Frequently the obstruction consists of a piece of turnip, mangel wurzel, carrot, or potato ; and sometimes a whole egg given by an ignorant attendant, with the erroneous view of curing colic, proves to be the offending agent. Sometimes balls, made of larger size than they should be, will not pass down the gullet, and becoming lodged there, cause choking. Animals with voracious appetites, writes Percivall, are especially apt to bolt their corn, gulping it down so rapidly that the successive portions, instead of passing into the stomach, accumulate in the gullet and block up its channel. Only a small collection, or a large one, may thus be made, before the animal manifests any uneasiness. All at once he leaves off feeding. He makes every effort to empty his gullet, and to relieve himself of his increasing distress. Should he not succeed, his throat and neck become, through his ineffectual exertions, spasmodically drawn up. Probably he gives every now and again a loud shriek, no less expressive of his own anguish, than excitive of the compassion of those around him. Should he attempt to swallow water, the fluid, together with the saliva abounding in his mouth, returns through his nostrils. These urgent symptoms are not, however, always present, and they depend chiefly on the position of the obstructing body. Thus, when it is in the upper part of the gullet, the distress, coughing, and slavering are very urgent. When the obstruction is in the neck, there is a visible enlargement in the course of the gullet, the general symptoms being great anxiety of countenance, sunken head, tremors, and partial sweats over the body, with great exhaustion, shortly after the occurrence of the accident.

The term choking is sometimes also employed for obstruction to the windpipe, which sometimes is pressed upon by a too small collar, or in other ways ; but it should be merely used to designate impaction of material in the gullet. When the portion of the gullet in the chest is obstructed, the symptoms manifested are usually not so severe. Sometimes even, the horse will drink a little water. Vomiting is uncommon in the horse ; and, when it does occur, the contents of the gullet usually escape through the nostrils, though at times they make their way through the mouth. When the whole length of the gullet is obstructed, the symptoms are most severe, and the danger necessarily greater. A condition which may be mistaken for choking is hellebore-poisoning, a case of which we have just recorded in the April numbers of *The Veterinarian* and *The Veterinary Journal* for the year 1886. On March 6, we were called to a heavy draught horse said to be choking. The symptoms observed by the owner had supervened three hours after the administration of a ball, containing a large quantity of hellebore. The animal was found to be retching continually, but although vomition does sometimes occur in such cases, it did not actually take place. The pulse was very irregular and feeble. The symptoms had been gradually becoming more severe, until, when death seemed imminent, advice was sought. Three ounces of whisky, with three ounces of aromatic spirit of ammonia, were ordered to be given every hour for six times, and then every two hours. In twelve hours' time the animal began to improve. On the following day he was much better, and tonics were substituted for the stimulants. The animal then made a rapid recovery. It may be needless to add that, on our first seeing the animal, nothing whatever was said concerning the poison which had been given with the idea of curing the "grease," from which the horse was suffering.

Sometimes professional men are called to cases of sore throat, which on examination prove to be uncomplicated instances of choking. Mr. King was called to a horse which was said to have a sore throat. The gullet had no impaction in that portion which is outside the chest, but all liquids taken were returned, the horse being quite incapable of swallowing them. The animal died, and in that portion of the gullet in the chest, a ball made of the ashes of tobacco was found. As in the case of hellebore-poisoning, the attendant did not mention having given anything, and so the probang, which would have been passed and saved the animal's life, was not employed.

In those cases where the impacted material is within reach, it may be removed by the hand through the mouth. If it is almost, but not quite, within reach it may be gently pressed upwards by an assistant, while the operator grasps it by his hand in the mouth. The tongue may also be drawn forward out of the mouth, as this will help in the upward movement of the foreign body. If an egg is lodged in the gullet, it may be broken by pressure if a thick needle is first passed through it. When we are unable to remove the body through the mouth, we may endeavour to move it up and down gently, and if this be successful, it is highly probable that the animal will then be able to swallow it. If this does not prove availing, and in all cases where the obstruction is due to dry food impacted in the channel, it is advisable to administer frequent draughts of a mixture of oil and water, not restraining the animal from regurgitating it again at will. The impacted food in the interval of giving the liquid may be gently manipulated up and down with the view of breaking up the mass. Should this not prove successful in relieving the animal, and if the matter can be felt from the outside, we must, nevertheless, not use the probang, as it generally in such cases merely hardens the dry food into a more compact lump.

In a case of choking by locust-beans, bran, and chaff, in which the symptoms were very severe, Williams administered water until the gullet was full of water. This induced a violent fit of coughing. The whole of the fluid was thus forcibly ejected, along with some of the impacted mass. The process was repeated after short intervals of rest. Each fit of coughing brought up more and more of the mass, until it was entirely expelled. The water was given through the nostrils, owing to the restiveness of the animal. When the obstruction consists of a piece of carrot or other solid body felt from the outside, it may be cut down upon and extracted. The wound is then sewn up by a few stitches of carbolized twine, and is afterwards dressed with antiseptic applications. Food must be withheld for a time after this operation, and at first should consist of linseed or oatmeal gruel. Indeed, in any case after the relief of choking, the animal should be fed on moist food, in order to allow the distended tube to regain its normal shape. The passage of a probang down the gullet requires great skill.

OSTEOPOROSIS.

WE shall have very little to say of osteoporosis or big head, as it is by no means a common disease. This affection of the bones is generally seen in young animals. The bones being ill-nourished, become light and of a spongy texture. Those of the face in particular are more especially affected, and the face thus becomes much enlarged, and altered in contour. This disfigurement of the features gives the animal a very peculiar appearance. Death generally ensues as the result of the diseased condition of the bones, but recovery sometimes follows judicious treatment. Laxative diet and the administration of vegetable and mineral tonics are necessary.

MELANOSIS.

MELANOSIS is an affection almost entirely confined to old grey horses. It consists in the growth of darkly coloured sarcomatous tumours in different parts of the body. The internal organs are often invaded by these malignant growths, but, as a rule, they just make their appearance on the tail, or on the region of the throat and neck, or on other parts. Sometimes a tumour will grow on old grey horses very gradually for several years, probably seven or eight or more, without producing any noticeable ill effects. It may then suddenly start growing more rapidly, and unless removed it may increase until it bursts, and forms an unhealthy ulcerating wound, which will not heal. It is commonly supposed that, if a melanotic tumour be removed, it will necessarily reappear.

In the year 1884, we removed a melanotic tumour from the throat of an aged grey pony. Some years previously, the owner was advised that recovery was impossible, as the tumour would grow again, and that the wound made would not heal. It, however, was increasing rapidly when we were called in. We advised its removal, and after putting the animal under ether and chloroform, which were used instead of pure chloroform, as the poor creature was broken-winded, the tumour was successfully taken away. The animal made a perfect recovery, and up to the present time there has been no recurrence of the growth.

CHAPTER IX.

OPERATIONS.

Administration of Anæsthetics. Firing. Bleeding.

ADMINISTRATION OF ANÆSTHETICS.

BEFORE performing a serious operation, it is generally customary to put the horse under the influence of an anæsthetic. Chloroform is the agent we almost exclusively employ for this purpose. Contrary to a notion somewhat widely spread, we may state emphatically that when administered carefully, there is not the slightest risk attending its use in healthy horses. The late Mr. D. Gresswell employed this agent for producing anæsthesia for many years in almost all serious operations on the horse in his extensive practice, and never saw any ill effect attending or following its administration. We have likewise used it very extensively, and have never known of any untoward results caused by its inhalation. During the administration, however, it is necessary to feel the pulse at intervals; for chloroform has a tendency to reduce the tension in the blood vessels, while ether has no appreciable effect of this kind. Three to five ounces of chloroform are generally effectual in causing insensibility. It is our practice to pour two ounces of chloroform into Gresswell's chloroform cap, an apparatus of which we append an illustration, and then to adjust it over the mouth of the animal.

We add amounts of about half an ounce at intervals, as may be necessary. If a horse struggles much, he generally requires a larger amount than if he breathes slowly and quietly. In some cases, we have found it necessary to give as much as six ounces before anæsthesia was complete. The idea of the danger of giving chloroform to the horse has doubtless arisen from the results of the wrong modes in which it is sometimes administered. Practically, the method we recommend will prove thoroughly reliable. An idea of the degree of insensibility produced by the chloroform inhaled may be gained by observing whether the eye is sensitive to the touch, or does not respond to the irritation caused when anything is brought into contact with it. When the operation is over, and the shackles with which the animal has been cast are loosened, it is well to give a good bran mash; but it is not advisable to dash water over him, or otherwise unnecessarily annoy and irritate him with no possible object. Should suffocation threaten, and the pulse show any sign of failing, the cap should be at once removed, and cold water dashed over the head.

D. Gresswell's Chloroform Cap.

FIRING.

WE have already incidentally spoken of firing in the treatment of several bony growths. We have now to speak of it more in detail. Before performing this operation, it is necessary to clip away the hair closely, from the part to be cauterised. In all instances where it is purposed to fire with the stripe-iron, it will be necessary to cast the animal, and then to administer chloroform, before proceeding with the operation. Of course, when it is only intended to fire with the prick-iron for a splint or spavin, casting and the administration of chloroform will not be necessary, as this operation is soon over, and causes but little pain.

Firing is a much more important operation than is generally believed, and requires infinitely more time, care, and judgment in carrying out thoroughly, than many people have any conception of. It is a most valuable counter-irritant, and is frequently productive of the best results. It is not only unadvisable, but it is absolutely inhuman, to withhold the administration of chloroform, in cases where it is intended to fire thoroughly. We may repeat again that, although we have employed chloroform very extensively indeed, we have never seen any ill effects following its use. The owner need not therefore fear to order its administration.

Prick-firing is especially adapted for side bones, ring-bone, bone-spavin, and splints ; whereas stripe-firing is especially useful in treating sprains of the back tendons and sesamoid ligaments, curbs, thoroughpins, windgalls, and bog-spavins. In firing for sprain of the back tendons and sesamoid ligaments of the fore leg, about ten strokes, with an interval of an inch between each, may be made obliquely from the fetlock upwards. Each stroke made with a stripe-iron should be gone over at least twenty times, the metal being heated to redness, and slightly cooled. This operation necessarily takes time and care. On no account should the skin be cut through. After firing, the animal's head must be tied up for a week, and in many instances it is advisable to blister the cauterised limb with equal parts of ointment of cantharides and of biniodide of mercury. The animal should be fed on mashes during this time, and should be led out daily for five or ten minutes. If a very severe action is not desired, we may fill up the burnt lines with Stockholm tar, instead of blistering. At the expiration of a fortnight, the incisions should be again filled with tar or grease. We have no hesitation in stating, that judicious firing is frequently the most efficient treatment for sprained tendons and sesamoid ligaments, curbs, windgalls, bog-spavins, and thoroughpins.

In firing the hock, it is necessary to exercise care to avoid the vein on the inside. Moreover, in firing in this part, the iron must not be too hot in operating on a thoroughbred horse, as otherwise it will penetrate through the skin, and cause an ugly gap. Oblique stripe firing is always attended with better results, than when this operation is performed in vertical lines. We may conclude our remarks on firing, by stating, that having a very large amount to do, we employ irons of our own patterns. The prick of the prick-iron we use is not more than from a fifth to a quarter of an inch in length, and our stripe-irons are not so bulky as those commonly in use.

BLEEDING.

WE may now say a few words regarding the practice of bleeding. Although scientific men are in the habit of inveighing, and with justice, against the absurdities which fashion imposes on its votaries in the matter of dress and various other customs, for instance, that most absurd custom of habitually taking certain noxious drugs, such as chloral hydrate, opium, tobacco, and large quantities of alcohol, still they themselves are not free from charges of worshipping at the same shrine. There is, strange to say, such a thing as fashion in medicine and surgery. At one time a particular drug comes into fashion. It is the custom to prescribe it, and this may be sometimes done, when it really is not needed. Bleeding, a useful practice extensively employed in former days, and perhaps too much so, is now, on the other hand, scarcely practised at all by some. It is most unfortunate that there should be this tendency to indulge in the freaks of fashion. There is no doubt that, just as in the past, some asthenic individuals, both men and horses and other animals, have been simply killed by excessive or misapplied bleeding ; so in the present, many cases of acute inflammation of sthenic type in plethoric animals have been lost, simply through lack of that abstraction of blood, which is so extremely useful. There are annually in a large practice, many animals, horses, beasts, sheep, in which there would not be the least chance of recovery, unless depletion of the excessive amount of

P

blood, which is circulating through the inflamed tissues, was carried out. Of course there are drugs which may to some extent serve a similar purpose ; but there are many instances, in which these do not prove so effectual, as the process of bleeding itself.

In treating plethoric, highly-fed animals, suffering from congestion of the lungs, acute laminitis or founder, lymphangitis or weed, inflammation of the lungs, and inflammation of the brain, bleeding is a necessary operation. We do not mention here special directions as to the method of bleeding. Under no circumstances should the amateur perform an operation fraught with so much possible danger. In some instances under our notice amateurs have used the lancet in opening the vein. This is very reprehensible. In bleeding it is our custom to use the fleam, opening the left jugular vein.

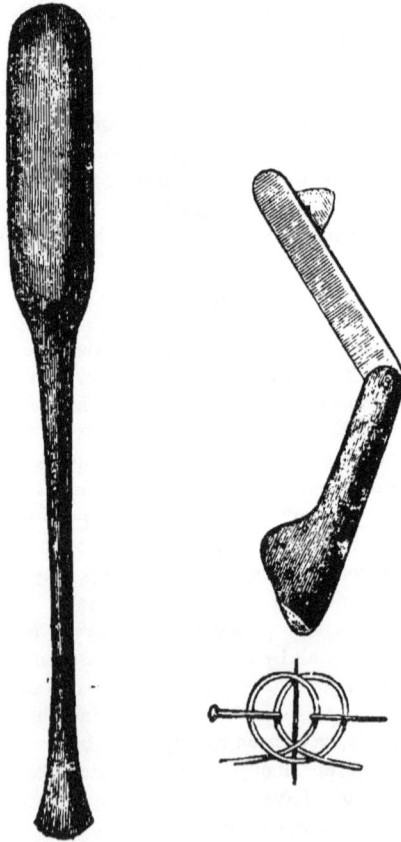

1. Bloodstick.   2. Fleam.   3. Method of closing the wound with a pin.

The course of the vein may easily be determined by pressing upon it, when the part furthest from the heart will become distended, by the blood accumulating in its channel. The spot we select for opening the vein, is nine or ten inches from the angle of the jawbone, and the amount of blood to be abstracted varies from one to four quarts. It is best to press the edge of the receiving vessel against the cut end of the lower part of the incised vein. When sufficient blood is removed, a pin is passed through the edges of the incised wound. Horse-hair is then wound round the ends of the pin in the form of a figure 8. In six days or so, the pin may be removed.

CHAPTER X.

DISEASES OF THE EYE.

Simple Ophthalmia. Recurrent or Periodic Ophthalmia. Amaurosis and Cataract.

SIMPLE OPHTHALMIA.

THERE are four diseases of the eye of which we must speak separately. These are simple and recurrent ophthalmia, amaurosis, and cataract.

Simple ophthalmia, is a disease of the eye characterised by inflammation of the white covering, and of the lining of the lids which is continuous with it. Together, the white and its continuation lining the inner surface of the lids are spoken of as the conjunctiva. As a rule, this inflammation is set up by an injury, or by the presence of a foreign body in the eye. We have already, in treating of pink-eye, mentioned that inflammation of the conjunctiva, is a feature characterising this variety of influenza. In ordinary cases of influenza, and other fevers also, the conjunctiva is frequently more or less inflamed. This inflammation may also have its origin in a cold, as is also very commonly the case in man.

Simple ophthalmia is easily known by the swollen condition of the lids, which are often wholly or partially closed, by the constant shedding of tears, by the projection of the haw or membrana nictitans, by the red hue of the inflamed part, and by the drawing back or retraction of the eye into its orbit. The cornea assumes a clouded appearance, but is only superficially inflamed. Sometimes, on examination, a foreign body will be found, and in such a case it is of primary importance that this should be removed carefully. The eye should be bathed with tepid water, and the animal should have a dose of aloes, and be fed for a time upon laxative diet, if the inflammation be at all severe. Locally, a few drops of a lotion made of two grains of sulphate of atropine to an ounce of water, should be dropped into the corner of the eye three times daily; and the animal, if very intolerant of light, should be placed in a darkened box. In two or three days, a lotion made of four grains of boric acid and ten drops of tincture of opium to each ounce of water, will prove valuable in restoring strength to the weakened eyes.

RECURRENT OR PERIODIC OPHTHALMIA.

RECURRENT or periodic ophthalmia is a more serious disease than the one we have just been considering. It is fortunately rather rare, but we have had several cases under treatment of late. These, fortunately, have all done well. Recurrent ophthalmia depends upon constitutional disturbance, induced by malhygienic conditions. Heredity is also a potent factor in predisposing to this disease. One should always bear this in mind, as no animal subject to it should be used for stud purposes. It has been noticed that animals bred in low-lying damp districts, are more liable to contract recurrent ophthalmia, than those living in healthy well-drained districts.

As a rule, this disease attacks one eye only, and the pupil of the disordered visual organ at first is seen to be contracted. It usually comes on very suddenly. There is no difficulty in distinguishing it from simple ophthalmia, to which affection it bears a general resemblance, from the fact that the whole eye being involved in the recurrent form, the inner parts assume a dull yellowish clouded aspect. An attack of recurrent ophthalmia runs through its acute stage in about eight days. The inflammation then abates somewhat, and the animal becomes a little more tolerant of light ; but a relapse, even when the best therapeutic measures have been adopted, will nevertheless sometimes occur, and the eye becomes almost or quite as bad as before. In the general way, about a couple of months elapses between each attack, but in the interval the eye is not restored to its normal condition. Unless cured, the relapses become frequent, and the disorganisation of the eye becomes so complete, as to cause total blindness. The first attack of recurrent ophthalmia is generally the most severe, and its symptoms the most intense. The eyelids become very inflamed and red, and the animal cannot bear to open them, or allow his head to be touched.

Early and judicious treatment is absolutely essential in treating recurrent ophthalmia. A dose of aloes should be given in the first instance, and the animal should be confined in a large, darkened, loose airy box, and fed upon laxative diet. Internally, a drench, containing liquor ammonii acetatis four ounces, bicarbonate of potassium half an ounce, and spirit of nitrous ether one ounce, may be given in several ounces of water three times daily. Locally, a few drops of a lotion of four grains of sulphate of atropine to an ounce of water, should be placed in the corner of the affected eye, three times daily. In the region of the head, behind the ear, a seton should be placed ; or, if this be not done, the same part may be thoroughly blistered with a mixture of ointment of cantharides and of biniodide of mercury. The food should be of a more nutritious kind, after the acute stage is over, and the drenches may then also be discontinued. These measures are generally effectual in curing the complaint. Should a relapse occur, the same steps must be repeated. The treatment sometimes carried out in this affection is very strange, and founded on no scientific principles.

AMAUROSIS AND CATARACT.

By amaurosis, we understand an affection of the eye in which the organ assumes a glassy appearance. It is frequently caused by derangement of the optic nerve, which expands to form the retina or nervous layer of the eye, and generally involves both visual organs at the same time. The pupil does not react to light, that is, it does not contract and dilate, when light is admitted into, and shut out from, the eye. The lids are wide open; and, when both eyes are affected, the animal is quite blind. This affection, when dependent upon disease of the nerve itself, is incurable; but, when it depends upon other causes, the eye may possibly be restored again to its normal condition. Therapeutic measures will usually avail nothing for the cure of this affection. Among the actual causes of this disease, are the growth of tumours in the brain, and blows on the eye.

By cataract, we understand an opaque condition of the lens of the eye, or of its capsule, which obscures the vision of the animal. It necessarily constitutes unsoundness, for, although it may be very small, it is nevertheless very liable to increase, and eventually results in blindness of the affected eye. It must be carefully distinguished from specks on the cornea, which are, as a rule, the result of previous inflammation.

FINIS.

INDEX.

—o—

A.

C

D.

G.

H.

J.

K.

L.

M.

Q.

R.

S.

Q

T.

THE YORKSHIRE POST,

The leading Conservative Organ in the North of England,

PUBLISHED DAILY, PRICE ONE PENNY,

Has a larger circulation than any other Newspaper in the Provinces, and circulates throughout Yorkshire, Lancashire, Lincolnshire, Nottinghamshire, and Durham.

About **3,000** *copies are sent daily through the post to the nobility, gentry, landowners, farmers, and professional men in the counties mentioned.*

No efforts are spared to obtain the latest and most reliable information in every department of news.

A special staff of reporters is maintained in the Houses of Parliament, for the purpose of giving full and complete reports of the proceedings, which are transmitted over two private wires, connecting the Branch Office in London with the Head Office.

The commercial intelligence is unsurpassed for fullness and accuracy.

Great care and attention is paid to sports of every description, and the paper is considered the best authority in the North of England on cricket, football, lawn tennis, and racing.

TERMS OF SUBSCRIPTION FOR PAPERS SENT BY POST, 10/6 PER QUARTER, PAYABLE IN ADVANCE.

UNEQUALLED AS AN ADVERTISING MEDIUM
In the districts in which it circulates.

HEAD OFFICE :—23, ALBION STREET, LEEDS.

BRANCH OFFICE :—80, FLEET STREET, LONDON.

BRANCH OFFICE :—73, MYTONGATE, HULL.

THE

YORKSHIRE WEEKLY POST,

AN EXCELLENT FAMILY NEWSPAPER,

(With Illustrations,)

PUBLISHED EVERY SATURDAY, PRICE ONE PENNY,

CONTAINS :—

Serial Stories by the best authors of the day ;

Special Articles on Social and Scientific Subjects ;

Echoes of Fashion ;

Angling Notes, written in a humorous and interesting style;

Extracts from the best Magazines ;

American Wit and Humour ; Household Recipes ;

Archæological Studies ;

Walnuts and Wine,—Gossip on Current Events ;

And Comprehensive Summary of the Week's News.

SENT BY POST TO ANY ADDRESS FOR 3/6 PER HALF-YEAR, PAYABLE

IN ADVANCE.

Works by Messrs. Gresswell.

"A Manual of the Theory and Practice of Equine Medicine."
10s. 6d.   London: BAILLIÈRE, TINDALL & COX, 1885.

"The Equine Hospital Prescriber."
2s. 6d.   London: BAILLIÈRE, TINDALL & COX, 1886.

"Veterinary Pharmacology and Therapeutics."
5s.   London: H. K. LEWIS, 1885.

"Analysis of Waterland on the Eucharist,"
By the Rev. HENRY WILLIAM GRESSWELL, M.A., Oxon.
1s.   London: JAMES NISBET & CO., 1886.

"The Veterinary Pharmacopœia, Materia Medica and Therapeutics."
London: BAILLIÈRE, TINDALL & COX.   (In the press.)

"The Wonderland of Evolution."
3s. 6d.   London: FIELD & TUER, 1884.

"The Evolution Hypothesis."
Capetown: Messrs. DARTER BROTHERS & WALTON, 1885.

"Some Pathological Bearings of Darwinism."
1s.   By Dr. D. ASTLEY GRESSWELL, B.M., Oxon., 1885.

"A Treatise on Human Therapeutics."
(Shortly.)   By Dr. ALBERT GRESSWELL, B.M., Oxon.